D. M. SHAPIRO

31 January 1998

ex dono Jacob

22.50

# Peasant Struggles, Land Reforms and Social Change

# Peasant Struggles, Land Reforms and Social Change

## Malabar, 1836–1982

P. Radhakrishnan

Sage Publications
New Delhi/Newbury Park/London

*First published in 1989 by*

**Sage Publications India Pvt Ltd**
M-32 Market, Greater Kailash I
New Delhi 110 048

**Sage Publications Inc**
2111 West Hillcrest Drive
Newbury Park, California 91320

**Sage Publications Ltd**
28 Banner Street
London EC1Y 8QE

Published by Tejeshwar Singh for Sage Publications India Pvt Ltd, Phototypeset by South End Typographics, Pondicherry, and printed at Chaman Offset Printers, Delhi.

**Library of Congress Cataloging-in-Publication Data**

Radhakrishnan. P., 1949–
Peasant struggles, land reforms, and social change: Malabar, 1836-1982 /
P. Radhakrishnan.
p. cm.
Revision of author's thesis.
Bibliography: p.
Includes index.

1. Farm tenancy — India — Malabar Coast — History. 2. Land tenure — India — Malabar Coast — History. 3. Peasantry — India — Malabar Coast — History. 4. Malabar Coast (India) — Social conditions. I. Title.
HD1511. 142M357 1989 333.3′ 1′ 5483 — dc19 88-35302

**ISBN** 0-8039-9593-8 (US)
81-7036-133-8 (India)

# Contents

## List of Tables

# List of Illustrations

# Acknowledgements

This book is an extensively revised version of my Ph.D. thesis on 'Land Reforms and Social Change'. From the time I commenced work on the thesis in 1975 right up to the completion of this book, I have had the help and encouragement of several persons. While I am indebted to all of them, I ought to mention at least a few names.

C. Rajagopalan, my Ph.D. supervisor and Professor and Head of the sociology department at Bangalore University, introduced me to social science research. But for his kind disposition and keen interest in my work, it would have been far more difficult for me to get a foothold in academics. N. Jayaram, Associate Professor in the same department, helped me in several ways, especially by offering valuable suggestions in the course of my research. K.K.N. Kurup, Professor in the history department of Calicut University introduced me to the village selected for special investigation, and allowed me to consult his personal collections.

Among my respondents who made my fieldwork possible A.N. Kotakkat needs special mention. He arranged for my stay in the village throughout the period of my fieldwork. He also helped me to establish contacts with peasant and political leaders in north Malabar.

On the home front I could always count on all, especially my brother P. Gangadharan and brother-in-law C.N.C. Unni.

A part of my thesis was first revised for publication as four self-contained articles in the *Economic and Political Weekly* issues of 13 December 1980, 26 December 1981, 25 September 1982, and 24-31 December 1983. The promptness with which Krishna Raj published these articles was itself very encouraging.

The second and more extensive revision of my thesis was during 1986–88, specifically for the preparation of the typescript of this book. It was facilitated by my work-place, namely, the Madras Institute of Development Studies (MIDS). The MIDS also helped me with a small financial grant for travel and additional data collection.

Within the MIDS, many of its members offered me generous help and encouragement: C.T. Kurien and K. Nagaraj commented on more than one draft of the typescript of the book. S. Guhan commented on the first draft. C. Guilmoto, S. Janakarajan and M.S.S. Pandian commented on part of the first draft. Guilmoto also helped me on the computer. C. Annadurai and S. Subramanian evinced keen interest in the progress of the book. Subramanian also helped me in several other ways. N. Chandrasekaran prepared the maps. S. Subbalakshmi, the MIDS librarian, helped me in getting rare books from other libraries. C. Narasimhan and J. Robinson word-processed part of the first draft. G. Dayalan, V. Mohan, and M. Suvakkin helped in xeroxing.

Apart from the MIDS, officials of the Tamil Nadu Archives helped me in consulting some of the rare documents and reports. Officials of the *Mathrubhumi Daily* at Calicut allowed me to consult their back issues. K.K.N. Kurup was again a source of immense help. He commented on the first draft and also arranged for my stay at Calicut for about two months. V.N. Deshpande, M.A. Kalam, and K. Venugopal were sources of considerable encouragement. The comments of the anonymous referee of my publishers helped in improving the typescript.

October 1988 P. RADHAKRISHNAN

# 1

# Introduction

The emphasis laid on land reforms today and the concern voiced when their progress is slow or unsatisfactory tend to give the impression that these are programmes undertaken mainly on governmental initiative. The history of land reforms in different parts of the country, however, shows that far from being liberal gifts from enlightened governments these have been historical processes necessitated by protracted peasant struggles.

The peasant struggles bear close similarity to the struggles which the industrial workers had to wage, particularly in Britain at the beginning of the industrial revolution. The long process by which numerous pieces of industrial legislations were enacted with a view to protecting the interests of the industrial workers, and the trade union was eventually recognised as a legitimate organisation, finds its analogue in the history of peasant struggles and land reforms. This suggests that no major legislation designed to protect and promote the interests of the oppressed and exploited groups can be ordinarily expected in the absence of persistent demands and protracted struggles; that governmental attempts at suitable legislations to enforce even the latest reforms are no more than belated responses to long-standing demands; and that in the ultimate analysis even the so-called planned changes of the present are undertaken not so much on governmental initiative as on the insistence of the oppressed and exploited groups. These issues are examined in this study in the context of the historical processes of peasant struggles and land reforms in Malabar and as an attempt at providing a new perspective to the analysis and understanding of rural social change on the basis of land reforms. Of these two dimensions of the study, the latter calls for some explanation.

In agrarian societies land is the most important means of wealth, and source of power and prestige, and rights in land are often hereditary with power and prestige tending to be ascriptive attributes. Consequently, changes in the institutional framework of agriculture, the pattern of ownership, control, and use of land, or more briefly, the land system, will have far reaching effects on the land-based social order in such societies.

Changes in the land system are occasioned by various factors, such as, territorial conquests, immigration, and extension of the economic frontier. In recent times, however, land reforms have been recognised as the most important of such factors. Thus Neale's apt remark that 'land reform is the chant of the times' and that 'in the present struggle of former colonial areas to raise their standards of living it has become axiomatic that land reform is a prerequisite to any advance, while the present tenurial systems are often blamed on the colonial powers' (Neale, 1962:3). In the two decades following the Second World War many of these colonial (and predominantly agrarian) areas achieved political freedom and formed national governments. It was when these governments embarked on development programmes that land reforms began to assume the importance of an 'explicit and strategic developmental issue' (Dorner, 1972:21). This was mainly due to the mounting pressures from the peasantry, which brought to the fore the injustices and inadequacies of the traditional land system and led to their recognition as major constraints on agricultural growth and social justice. In this connection it is relevant to note the following observations by Joshi.

> The anti-colonial awakening in these countries was now transformed into an urge for rapid economic development and social progress. The national leaders in these countries were in search of new paths to prosperity for the people through an effective use of the newly-won state power. Since the peasantry constituted a major anti-colonial force in most of these countries in the period of the nationalist struggle, land reforms for the emancipation of the peasantry from feudal burdens became a necessary part of the process of national planning for economic development. (Joshi, 1976:36)

In India, also, land reforms began to assume importance immediately after independence and have figured prominently in the Five Year Plans. In fact, the initial enthusiasm of the Government of India was of such a high pitch that Thorner rightly commented as early as in the 1950s that 'the sum total of land reform laws enacted in India since 1947 constitutes, I believe, the largest body of agrarian legislation to have been passed in so brief a span of years in any country whose history has been recorded' (Thorner, 1976:18).

Briefly, the land reforms envisaged in India during the last three decades have involved the abolition of intermediary tenures, regulation of rent and tenancy rights, conferment of ownership rights on tenants, imposition of ceiling on holdings, distribution of surplus land among the landless and the land-poor, and the consolidation of holdings.

Although none of the states have succeeded in implementing all these reforms, even the limited implementation of some of them has had tremendous effects on the traditional land system and through it on the land-based social order. Generally, such implementation has both reduced in importance and transformed in quality the very top of the traditional agrarian hierarchy and eliminated a peculiar form of land ownership and many of the elements of a 'feudal' society (Beteille, 1974:78). Particularly, the abolition of zamindari has made the land ownership pattern in ex-zamindari areas 'much more equitable and free from some of the worse features of feudal exploitation' (Dantwala, 1957:8), and greatly eroded the power of the zamindar class, resulting in the emergence of a 'class of substantial peasants and petty landlords . . . as the dominant political and economic group' (Bell, 1974:196). Even the unsuccessful measures have had some effect through the fear of their implementation inducing potential losers to sell or rearrange their holdings (Byres, 1974:224).

While such effects undoubtedly bring to the fore the great significance of land reforms as a potent force of social change, it is important to note that quite unlike the recent origin of the recognition of this significance, land reforms are by no means recent. On the contrary, they have been a historical process. It is also important to note that more often than not, even this recent recognition has not been spontaneous but has been brought to bear on the government by the cumulative effect of the earlier peasant struggles and the resultant land reforms. What is more, due to resistance by the landlords and the lack of political and bureaucratic will, even this forced recognition has seldom been the cause of effective land reforms unless it has been preceded, paralleled or followed by pressures from the peasantry. It is in this sense that the cause-effect relationship between peasant struggles and land reforms has to be understood, and even recent land reforms have to be viewed as only the latest phase of a long historical process of such reforms necessitated by the protracted struggles of the peasantry.

If land reforms are thus viewed, and if many of the recent changes in agrarian societies are attributed to such reforms, the causes of these changes will have to be finally traced to those social forces which necessitated the earlier reforms. This should mean that studies of the complex and dynamic processes of social change on the basis of land reforms, or for that matter any other social policy with a history of its own, cannot be merely confined to recent reforms and their most visible social effects, but, of necessity, will have to be historical and comparative. This study is in keeping with such a perspective.

Although there have already been several studies on land reforms, most of them have been concerned mainly with the relevance, nature, scope, chronology and evaluation of such reforms. Though the effects of land reforms on the land system and social order have also figured in some studies, only a few of them have treated land reforms as a basis for the analysis and understanding of rural social change. Even in these few studies, the land reforms treated as a basis for such analysis and understanding have been only the recent ones and there is hardly any study which has attempted a careful analysis of the historical process of land reforms in the context of the history of peasant struggles. As a matter of fact, most studies on land reforms have treated these reforms in isolation of peasant struggles, though, as noted by Alexandrov, 'it is impossible to discuss the dynamics of socio-economic and political change in the developing countries, without including a careful analysis of the history and progress of peasant movements' (Alexandrov, 1974:351). Similarly, most studies on peasant struggles have been concerned mainly with the nature and characteristics of these struggles to the exclusion of land reforms. It is to fill this lacuna by attempting to analyse and understand rural social change in the context of the historical processes of peasant struggles and land reforms that this study is undertaken.

Apart from being historical, the study is both macro and micro, albeit at the risk of an inevitable disjunct between the two. Against the backdrop of a brief sketch of the land and society in pre-British Malabar, at the macro-level the study will first describe the historical processes of peasant struggles and land reforms in Malabar from the early nineteenth century right up to 1970 when these processes culminated in the Kerala Land Reforms (Amendment) Act of 1969—the most drastic of all the hitherto land reforms enactments. This will be followed by an examination of the implementation of this Act at the state level, and its major effects on the land system. At the micro level the study will analyse the changes brought about by this Act in the land and society in a selected village. This micro-analysis is more in the nature of an appendix to the macro study and is intended to provide a general idea of the nature of the changes brought about by the Act which cannot be easily comprehended at the macro level.

The choice of Malabar for the study has to be explained by the importance of Kerala in general and Malabar in particular for studies on land reforms. As observed by Raj, Kerala is 'one of the few areas of India, where, for a long time, the relationship between landlords and those who functioned under them resembled (except for differences introduced by the caste system) that prevalent under feudalism in Europe' (in Varghese, 1970:vii). Prior to its reorganisation as a linguistic state in 1956, it had the highest percentage of area under tenancy in India, and was perhaps the only state characterised by such a bewildering variety of land tenures which were 'almost unique in respect of their complexity and multiplicity of incidence' (Oommen, 1975:8). It has a long history of land reforms dating to 1863 in Cochin, 1865 in Travancore, and 1887 in Malabar, and a longer history of peasant struggles as will be evident from this study. Finally, as observed by Nossiter 'in a national context Kerala's land reform, begun by the 1957–59 ministry, brought to legislative fruition by the UF ministry, and implemented by the Mini and Maxi fronts has attracted attention as the most radical, comprehensive, and far reaching in South Asia' (Nossiter, 1982:292). Although Kerala is now a single political unit, until its reorganisation as a linguistic state its three constituent regions formed three separate politico-administrative units, with strongly contrasted land systems characterised by peasant proprietorship in Travancore, absentee land ownership in Malabar, and a combination of both in Cochin. Of these three units, because of its peculiar land system, Malabar had the highest percentage of tenancy, the most complex land system, and the worst form of landlordism. And, as peasant struggles struck deeper roots in this region and continued to be more militant and protracted than in the other regions, it is only to be expected that the impact of the 1969 Act has been greater on this region than on the other two.

Within Malabar, the choice of the village selected for special investigation is also relevant for several reasons. The traditional society in this village, understood in terms of its caste system and the caste-based land system, had been a fair specimen of the larger society in Malabar. Its peasantry had been active participants in the third and most active phase of the peasant struggles of the region. Its record of the implementation of the 1969 Act closely conforms to the larger state pattern. Finally, precisely for these reasons, analysis of the changes in its land system and social order brought about by the implementation of this Act will help in visualising the changes in the land system and social order in the larger society.

As will be evident from the sources cited, the description in Chapter 2 of the land and society in pre-British Malabar is based on published historical accounts. The review of peasant struggles and land reforms in Chapters 3 and 4 is based on similar accounts, published works on peasant struggles and land reforms in Malabar, newspaper reports, biographies of and interviews with important peasant and political leaders of the region, and published and unpublished government documents. Chapter 5 on the implementation of the 1969 Act and its impact at the state level is based on administration reports of the Land Board and Land Tribunals and other published sources. Chapters 6 and 7 on the implementation of this Act at the village level are mainly based on data collected on two field visits, each of three months' duration.

# 2

# Land and Society in Pre-British Malabar

Before the British took over in the last decade of the eighteenth century, Malabar was not part of a centralised polity, or of an all-embracing political superstructure, as it turned out to be in subsequent years. Instead, it comprised numerous petty principalities as observed by Ibn Batuta in the fourteenth century (see, Lee, 1829:167), and by Barbosa about two centuries after him (see, Dames, 1921, Vol. 2:6). Probably Malabar was so divided from at least the tenth century A.D.

Barbosa began his description of the state and society in Malabar during his time with a local legend that not less than six hundred years before him all Malabar[1] was one kingdom of the mighty Cheraman Perumal, and that before he embraced Islam and left for Mecca, he divided it among his kinsfolk (see, Dames, 1921, Vol. 2:2–6). In his attempt to trace the history of land tenures in Malabar, Logan, one of the British

[1] That is, not merely what is now known as Malabar, namely, northern Kerala or the erstwhile British district of Malabar, but the whole Malayalam language area stretching about 350 miles from Cambola (Kumbla) on the Chandragiri river in south Kanara to Cape Comorin.

Collectors in Malabar who headed the controversial Malabar Special Commission of 1881–82 (of which more later) also recounted this legend (see, Government of Madras, 1882, Vol. 1:xi). Based on an analysis of four ancient deeds, that is, the three well-known deeds belonging to the Jews and Syrians of the Cochin and Travancore States and a stone inscription found in Calicut, Logan contended that at the head of all in the organisation of the State in the eighth and ninth centuries A.D. was the Perumal or Chakravarti, or the Emperor, and that curiously enough the very same chieftains with whom the British came into contact when assuming charge in the eighteenth century were in those days the *udayavar* of the *nads* into which the empire was divided. Based on an examination of the other historical material then available to him, he also made some other important observations. For instance, between the ninth century A.D. and the arrival of the Portuguese in the late fifteenth century A.D., the chieftains of the *nads* reigned almost unchanged within the limits in the *nads* which they held under the Emperor in the eighth and ninth centuries; that despite the further weakening of the intrinsically weak local governments due to the protracted and expensive wars with the advent of the Portuguese and the invasions by the rulers of Mysore, the constitution of Malabar society when the British took over was essentially the same as it had been for centuries; and that no explanation other than custom became king paramount seemed to account satisfactorily for the absence of any upheaval of society when the last Perumal disappeared (Government of Madras, 1882, Vol. 1:x–xii).

Whether Malabar's political divisions in pre-British times were long preceded by its existence as a centralised polity or part of it, and effected by the Perumal himself or whether they were characteristic of a segmentary state are still conjectures. What is relevant to note here is its persistence, despite all divisions, as an integrated socio-economic region, thanks to the prevalence of what Logan described as custom as king paramount, through the mutually reinforcing institutions of land and caste.

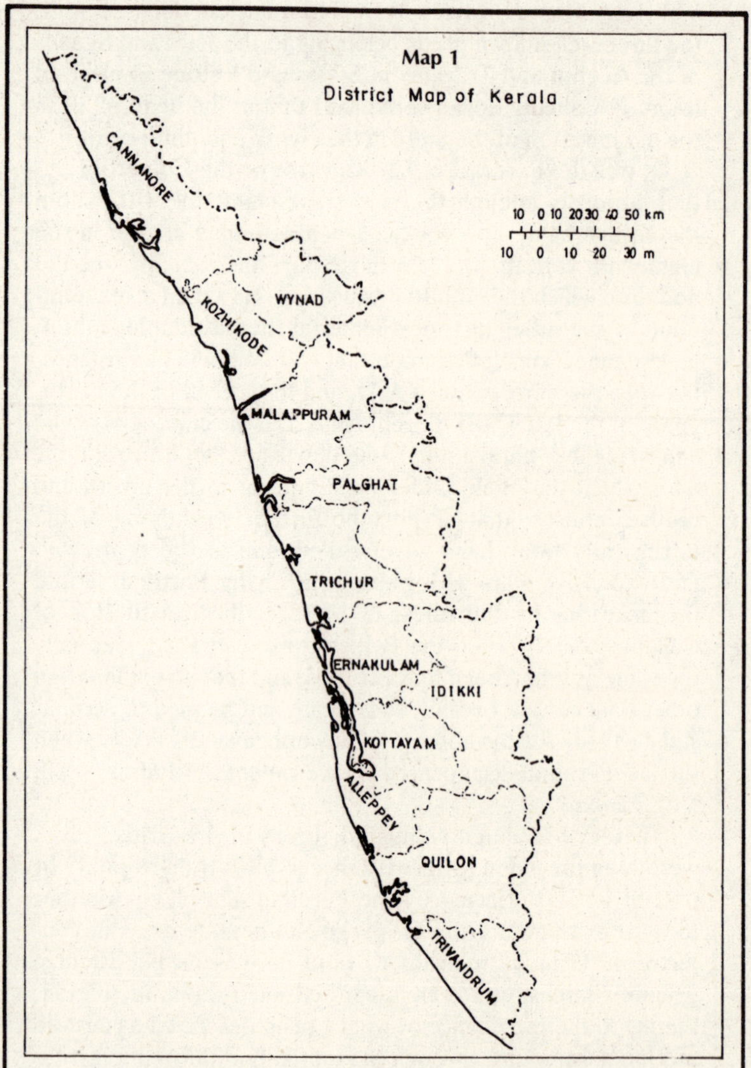


**Map 1**

District Map of Kerala

## Map 2

### Taluk Map of British Malabar

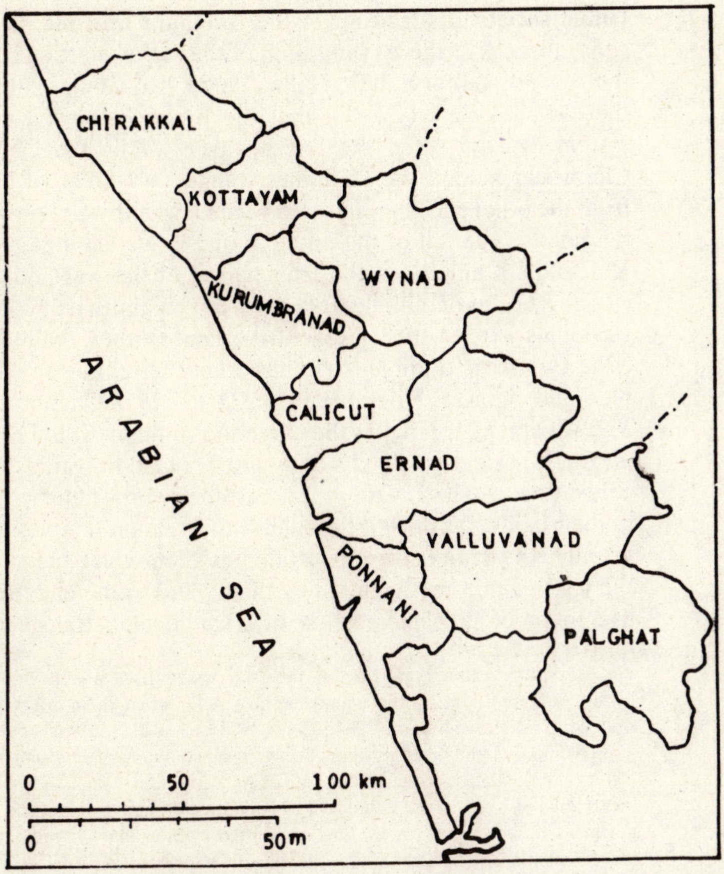

## Land System

From the available literature on the history of Kerala land tenures it is possible to discern three distinct interpretations of the caste-based, Brahmin-centred, and upper-caste-dominated land system in pre-British Malabar. The first is that it was a creation of the Aryan immigrants, and not in existence in earlier times. The arguments run thus. In Kerala as a whole, private property began to evolve even before the Sangam age. Until about the sixth century A.D., land was owned by those social groups which are now at the bottom of Hindu society. Its transfer to the Brahmins was one of the consequences of the Aryanisation of the region starting from the seventh century A.D. (Pilla, 1966:8–10). It was during the eleventh century A.D. that the Brahmin-centred land system evolved when the forces unleashed by the Chola-Chera war helped the Brahmins wangle vast areas of land from the other social groups and become wealthy and powerful *janmis*. The fall of the united polity in Kerala under the Kulasekhara empire in the later period of the war and the rise on its ruins of innumerable petty principalities helped the Brahmins extend their sway further, when they made the caste system[2] and the *janmi*-centred land system more complex and rigorous (Menon, 1964:325–69; 1978:Chapter 12; 1979:Chapter 2). Finally, the evolution of the new and complex land system involved a long series of land transfers. The cultivators who had more or less customary proprietary right in their lands transferred their lands to Brahmin temples and through the temples to the Brahmins themselves believing the legitimation-myth spread by them,[3] and to the chieftains as a token of allegiance. These new land owners transferred

[2] According to Sreedhara Menon, the caste system itself was introduced by the Brahmins during the eighth century A.D. when their influx had reached its peak. See Menon, 1964:325–69; 1978:Chapter 12; 1979:Chapter 2.

[3] Tradition, as embodied in works like *Keralolpatti* and *Keralamahatmyam*, has it that Kerala was raised from the sea by Parasurama, the axe-bearing incarnation of Vishnu, and gifted to the Nambutiris brought by him from the north, as an atonement for the sins committed by him, that for protecting these Brahmins, he brought Sudras, people of the lower social order, that he ordained rules of conduct and customs for them, and so on. See Innes, 1951:37–38; Logan, 1951a, Vol. 1:221–24.

their lands on *kanam* tenure to the Nayars. As the Nayars themselves were placed in the upper stratum of society and accepted the caste rigidities imposed by the Brahmins, like the latter they also spurned labour. Their holdings were therefore passed on to people of the lower strata of society on inferior tenures (Varghese, 1970:12–16).

The second interpretation is the legitimation-myth already mentioned, that Parasurama reclaimed Kerala from the sea and made it over to the Nambutiris in full proprietary right (*janmam*). As Logan rightly observed, though inconsistent with available historical facts, the unquestioned acceptance and adoption of this myth as a fundamental truth was an unfortunate mistake made at the outset of the British rule, and one of the consequences of the glamour which the Brahmins managed to cast over the early British administrators (see Government of Madras, 1882, Vol. 2:Chapter 2).

The third interpretation is by Logan himself. Attempted as early as 1881, it has remained the most exhaustive and systematic account of the history of the Malabar land system attempted hitherto, and a subject of serious administrative and academic discourse for more than a century now. Though it has been partly or wholly refuted by several British administrators, none of them could offer a plausible alternative in its historical context.

Logan argued that 'to attempt to write even a short account of the Land Tenures of Malabar without any preliminary reference to the political organization which gave them birth, would be as sensible a plan as to attempt to give an intelligible account of the law of real property in England without mentioning the feudal system' (Government of Madras, 1882, Vol. 2:3), and accordingly attempted to trace the history of the land question in Malabar from the earliest times down to the late nineteenth century in terms of its political organisation. Based on the material on the early history of Malabar collected by him, especially the four deeds mentioned earlier, he first showed that the organisation of the state in the eighth and ninth centuries A.D. was by guilds or corporate bodies, each, down to the lowest and the meanest, having its distinct function to perform in the body politic. At the head was the Perumal or Emperor. Below him came a number of chieftains of the *nads*, including the well-known Rajas of Venad (Travancore), Eralanad (Calicut), Valluvanad, and Nedumpuraiyurnad (Palghat). Each *nad* being a territorial organisation of the Nayars into Six Hundreds, who were *kanakkar*, overseers, or protectors of all within it, the chieftain was not an absolute ruler but simply the head of the guild, the executive agent of the Six Hundreds' wishes and their representative in all public affairs. Beneath the *nads* came other but subordinate corporate bodies each with a greater or smaller share of self-government:

> A certain class, the *Iluvar* or Tiyar (*Cingalese* or Islanders), were told off to plant up the waste lands and manage the gardens. Another class, the *Vellalar* (irrigators), were the cultivators of the wet lands . . . . [There] have survived, even to the present time, similar organisations of the handicraftsmen, artificers, etc. (astrologers, carpenters, goldsmiths, washermen, etc.) known as the *Cherujanmakkar*, or small birthright holders. (Government of Madras, 1882, Vol. 1:x–xi; Vol. 2:19)

The functions thus acquired in the body politic became birthrights, and the holders thereof were entitled for their services to certain portions of the produce of the lands within the limits of their birthrights. Logan argued that in a society thus organised, in short, like a well-ordered household, there could not have been ownership of land in the modern sense (Government of Madras, 1882, Vol. 1:xi; Vol. 2:19). Stated differently, in the land system of pre-British Malabar, characterised by what Logan also described as 'corporate unity' or 'joint proprietorship,' each of the principal land right holders, viz. the *janmi, kanakkaran*, and *verumpattakkaran*, was a part-proprietor, who, as a matter of customary right was entitled to one-third or an equal share of the net produce. The net produce was the share left over after providing for the *cherujanmakkar* or all the other birthright holders such as the village carpenter and goldsmith, and the agricultural labourers who helped to gather, prepare and store the produce. The distribution of the respective shares was done by the *kanakkaran*. In ancient times the Perumal might have also obtained a share of the produce. This is indicated by Logan's observation that 'the share of the produce left over after providing liberally for the cultivating castes was styled *Pattam*, that is, the *Pad* (authority's) share' and that 'this *Pattam* was divided equally, share and share alike, between the *Janmi* and the *Kanakkar* who had between them inherited the old Government land revenue of Malabar' (Government of Madras, 1882, Vol. 1:xxvi–vii; Vol. 2:23). However, the information given by him is too inadequate to arrive at any definite conclusion.

At the risk of simplification, the land system of Logan's description may be summarised as follows:

### *Janmam*

Highest in the land system was what has come to be known as *janmam*, the Sanskritised form of the Dravidian phrase *nir attiperu*, that is, birthright acquired by coming in contact with *nir* or water, or water-contact-birthright. The right conveyed by it was not a free-hold or an allodium, but an office or dignity with a fixed share of the produce of the soil. The Nambutiris who entered Malabar in large numbers and as an organised body in the seventh and eighth centuries A.D., and who were received with open arms by the Perumals, managed to obtain commanding influence in the region precisely at the time when the Perumals, threatened by the Western Chalukyas, sought succour from every quarter. This, and the fact that as a very ancient right in the soil *janmam* was in the sole gift of the Perumal, explain how the Nambutiris became large *janmis* even at an early stage of the evolution of the land system in Malabar. From the time the last Perumal left right up to the sixteenth century A.D., the chieftains who remained almost without any change of limits in their respective *nads* did not confer on themselves the prerogative of creating *janmam* holdings—a prerogative which had belonged to their emperor alone. But, the weakening of their power after the Portuguese advent led to the creation of numerous pseudo-*janmams* (Government of Madras, 1882, Vol. 1:xii–xv; Vol. 2:15–19). This might have further helped the Nambutiris emerge as the predominant group of *janmis*.

### *Kanam*

Beneath *janmam* was *kanam*, the right to supervise or protect all the inhabitants of the particular *nad* wherein the land lay. Though socially subordinate to the *janmi*, for this service the *kanakkaran* enjoyed a fixed share of the net produce of the soil equal in amount to that enjoyed by the *janmi*. In other words, like the *janmi*, the *kanakkaran* was also a part-proprietor of the soil to the extent of one-third of the net produce (including the right to sell, subdivide and sublet his holding). In theory, he was never a fixture on any particular piece of land, for, he was the hereditary protector of his *nad* and not, unless under special grants, the hereditary protector of any particular bit of it. In practice, however, he was permanently attached to his holding, as the *janmi* would not dare to oust him for fear of spreading insecurity and discontent among the *kanakkar* who remained. On this point Logan has quoted the views of Rickards, the second of the Principal Collectors of Malabar, that 'it was, indeed, the interest of the *Janmis* to treat them [*kanakkar*] well, for they were not only the chief means of providing the *Janmis*' income, but composed his retinue of armed followers, so essential to his consequence and weight in the aristocracy'. On his part, the *kanakkaran* was, however, free to transfer his services to any one else who suited him better. If he relinquished his holding he was to be paid for, or allowed to sell any improvements he made upon it and his position as a member of the protecting caste made it very difficult for the *janmi*-to refuse such payment (Government of Madras, 1882, Vol. 1:xxvii, xxx, xliv; Vol. 2:17, 29).

### *Kuzhikanam*

Next in importance to *kanam* was *kuzhikanam*, a tenure on freshly reclaimed lands. Its holder, the *kuzhikanakkaran*, enjoyed the whole of its produce for a certain period (probably not exceeding twelve years). Perhaps he paid a trifling fee of two fanams on entry, but that was 'more as an act of fealty than as recompense for the privilege of possession'. At the end of this period he was paid at customary rates for the improvements made by him, in terms of the shares due as *pattam*, before the customary law regulating the distribution of the produce came into operation (Government of Madras, 1882, Vol. 1:xxx–i). Logan's reference to this tenure is too brief. This was probably because of its interim nature. Once the sharing of the produce was begun, it automatically became one of the principal tenures, presumably the *kanam*.

### *Verumpattam*

The last tenure, *verumpattam*, was the holding of the actual cultivator under the *kanakkaran*. In many respects the *verumpattakkaran* was socially subordinate to the *janmi* and the *kanakkaran*. Notwithstanding that, he was also part-proprietor to the extent of one-third of the produce (including the right to sell, or sublet his holding, and to payment for improvements when he relinquished his holding). Probably, the *verumpattakkar* were originally drawn from the Nayars as *vellalars*, irrigators, or cultivators of wet land, in addition to their own peculiar function as *kanakkar*; and from the Tiyyas as cultivators of the waste and garden lands (Government of Madras, 1882, Vol. 1:xiii, xvi, xxxi–ii).

Though Logan's description of the three-tier land system clearly brings out his contention that in pre-British Malabar property rights were imperfectly developed and regulated by customary rights, it needs to be emphasised that it has certain limitations. It does not explain the coexistence of a two-tier land system as well, with *janmam* at the top and either only *kanam* or only *verumpattam* below it, as his data analysed in the following chapter seem to indicate. Nor does it explain the complexity of *kanam* and *verumpattam* rights, with each having a multiplicity of sub-tenures. More importantly, even in its imperfect form, whatever advantages the land system held out were virtually confined to the *janmis* and the *kanakkar* and did not percolate the other groups to any significant extent. In relation to these groups the land system was highly oppressive and exploitative—a point which Logan did not emphasise adequately. This was so for at least two reasons. First, the pyramidal structure of the land system with each *janmi* having a large number of *kanakkar* under him, and each *kanakkaran* having numerous *verumpattakkar*, apart from the numerous undertenants in the last two groups. In such a structure equal distribution of the shares of the produce to the three principal groups, as mentioned by Logan, would not have resulted in equality in the total produce received by each. The *janmi* received an equal share from each of his *kanakkar*. The *kanakkaran* received an equal share from each of his *verumpattakkar*. But the *verumpattakkaran* received only his single share.[4] Second, the privileges and prerogatives which the *janmi* and the *kanakkaran* enjoyed by virtue of their superior status in the land system apart from the status of the former as undoubtedly lord paramount over all the occupants of his *janmam* holdings, and of the latter as master of all the lower classes settled on his holdings. To refer only to the *janmi*, 'in addition to [the] share of the produce of the soil, the *Janmi* was also entitled to various ranks and dignities of sorts. He was sometimes commandant of the militia of the *Desam* (hamlet); he was also generally, perhaps invariably at first, a man of authority in the protector guild of *Nayars*; he was often the trustee of the village temples'. Because of these, and other privileges and prerogatives, 'the price of the empty name of *Janmi* was estimated as worth exactly half the price at which the productiveness of the land to the *Janmi* was valued' (Government of Madras, 1882, Vol. 1:xiv). The *janmi* could use these privileges and prerogatives as a means of social closure to perpetuate his social domination and the subordination of the groups below him, especially the lower ones. To cite just one instance as reported by Logan:

Persons [Hindus] who incur the displeasure of their

[4] The data collected by Logan reveal that in 1881–82 his sample of 110 *janmis* of different classes, together holding 14,567 pieces of land in the whole of Malabar, had under them as many as 3,706 intermediaries, and 14,034 actual cultivators. This means that for every *janmi* there were as many as 34 intermediaries and 128 actual cultivators. No doubt, these data represent a truncated land system, especially because of the widespread decline of *kanam* into *verumpattam* as will be discussed in the following chapter. Nevertheless, they do give some general idea of the number of shares which each *janmi* might have received, though it is difficult to hazard any such inference in the case of the *kanakkaran*.

> *Nambutiri Brahmin Janmis* are subjected to what is called *Desa Virodham*, i.e., the enmity of all residents of the *Desam* or hamlet, and *Svajana Virodham*, i.e., the enmity of their own caste-people. This excommunication entails many most unpleasant and unjust consequences. The excommunicated may not be shaved, nor their women purified after confinements; their presence in the temples and at the bathing tanks is stopped; they are prevented from taking water from wells, and others are prohibited from giving it to them; people dare not help them with their domestic ceremonies; all who do them service, or give them aid, or even sometimes are seen speaking to them are either themselves excommunicated or fined. The smallest show of independence is resented as a personal affront. (Government of Madras, 1882, Vol. 1:xxvi)

As it is only to be expected, this land system was intrinsically and inextricably bound up with the caste system. In fact, both existed and persisted on mutually reinforced legitimacy and support as will be evident from a brief discussion of the caste system.

## Caste System

Barbosa described 18 castes of natives in all Malabar, besides immigrant merchants and traders with customs of their own (Dames, 1921, Vol. 2:7, 70). These castes are listed in Table 2.1 in the order he assigned to them, along with their more recent names as given by Dames, his translator and editor. As observed by Dames, Barbosa's description, though not exhaustive, includes nearly all castes of importance in the whole of Kerala, and the correspondence is very close not only in name but in occupation and social precedence (Dames, 1921, Vol. 2:70).

### *Brahmins*

The Brahmins described by Barbosa are the Nambutiris, who ranked first in the traditional caste hierarchy in Malabar. As

**Table 2.1** ***Caste-Ranking in Pre-British Malabar***

| *Castes Described by Barbosa* | | *Corresponding Names Given by Dames* |
|---|---|---|
| 1 | *Higher Castes* | |
| 1.01 | Kings | Kshatriya, Samantan, Eradi, Vallodi |
| 1.02 | Bramenes | Nambutiri |
| 1.03 | Nayres | Nayar |
| 1.04 | Biabares | Vyapari, Ravari Nayar |
| 1.05 | Cuiavem | Kusavan |
| 1.06 | Mainato | Vannathamar |
| 1.07 | Caletis | Chaliyan |
| 2 | *Lower Castes* | |
| 2.01 | Tuia | Tiyan, Izhuvan |
| 2.02 | Manen | Mannan |
| 2.03 | Canaqua | Kaniyan, Kanisan |
| 2.04 | Ageres | Asari |
| 2.05 | Mogeres | Mukayar, Mukavar |
| 2.06 | Monquer | Mukkuvan |
| 2.07 | Betunes | Vettuvan |
| 2.08 | Paneens | Panan |
| 2.09 | Revoleen | Eravallen (Cheruma?) |
| 2.10 | Poleas | Pulayan |
| 2.11 | Pareens | Parayan |
| 3 | *Immigrants* | |
| 3.01 | Chatis | Chetty |
| 3.02 | Guzurates | Gujarati Banya |
| 3.03 | Mapuleres | Mappila, Mappilla, Moplah |
| 3.04 | Pardesi | Pardeṣi or 'foreign' Muhammadan |

**Note:** The appellations 'higher castes' and 'lower castes' are as given by Barbosa and Dames. As castes below the Nayars are not considered higher, it will be more appropriate to refer them as 'non-polluting' and 'polluting'.

**Source:** Reproduced from Dames, 1921, Vol. 2:71.

noted by Logan, they settled most largely in the Valluvanad, Ponnani, Ernad, and Calicut taluks of south Malabar, and sparsely in the Kurumbranad, Kottayam, and Chirakkal taluks of north Malabar (see Government of Madras, 1882, Vol. 2:11, 14–15).

In their socio-economic status the Nambutiris were a highly differentiated group (see Gough, 1972: 306). Despite this, and despite their small size, as a caste they were the most privileged and influential of all social groups. Their highest ritual ranking, hereditary land right as *janmis*, and patrilineal family organisation characterised by primogeniture were probably the most important reasons for this.

Dames noted that even in the first quarter of this century the Nambutiris were a power in the land, owing partly to their spiritual ascendancy, and partly to their position as owners of much of the soil of Malabar, and wielded a powerful conservative influence over members of the ruling castes and high class Nayars (Dames, 1921, Vol. 2:33). The extent of their power and influence in the pre-British times when Malabar was relatively more pristine is easily imagined.

Two immigrant Brahmin castes—the Pattars and the Embrantiris—were not discussed by Barbosa. These castes were also privileged as small *janmis* and as groups of high ritual ranking, though not so much as the Nambutiris.

### *Other High Castes*

Below these Brahmin castes ranked the lineages of the royalty and chiefs whom Barbosa described as kings and placed above the Nambutiris. Though too small in size with membership confined to a few matrilineages, they were also big *janmis*.

Below these lineages ranked the tiny groups of Ambalavasis whom Barbosa did not discuss. They did not have any direct right in land, and lived as temple functionaries, mostly in the Brahmin settlements.

Ranking below the Ambalavasis were the Nayars who accounted for the bulk of the non-polluting castes. Innes, one of the British Collectors in Malabar, noted that the term Nayars in its widest sense includes a very large number of castes which are now distinct, in fact, all between the Ambalavasis on the one hand, and the polluting castes on the other; and that at the end of the scale are castes whose traditional occupations are by no means military, though there is little doubt that the Nayars proper were primarily a military class, probably a sort of yeomen militia, who devoted themselves to agriculture in times of peace (Innes, 1951:115).

These various Nayar castes may be classified into two

broad groups of high-caste Nayars or Nayars proper, and low-caste Nayars. The former held the various hereditary offices, such as of retainers or vassals, attendants of the Nambutiris, princes, and chieftains, and writers of the kings who received the hereditary title 'Menon'. With their hold on these hereditary offices and on land, as both *janmis* and *kanakkar*, as a group they enjoyed a unique and lasting position in Malabar society. Their importance is also quite obvious from Logan's observation that 'the domain of a chieftain is usually styled as a territory of so many thousand Nayars, divided into so many Taras or desams, with so many commandants, and extending (from north to south) a distance of so many leagues' (Government of Madras, 1882, Vol. 2:4). Logan also described the Nayars as the 'protector guild' of Malabar and wrote that 'their functions in the body politic have been tersely described in their own tradition as "*the eye*"; "*the hand*"; and "*the order*",' (Logan, 1951, Vol. 1:v–vi).

As in the case of the Nambutiris, the family organisation of the Nayars was best suited to prevent fragmentation of their land holdings. But, unlike the former, they were matrilineal and lived in the impartible joint family called *taravad*.

The low-caste Nayars comprised the mercantile, artisan and service sections of the Nayars, the numerous non-polluting castes of merchants, oil-mongers, washermen, potters, barbers, weavers, and so on, some of whom also served as under-tenants of the Nambutiris and the Nayars. Probably, this group included the last four higher castes of Barbosa's description: Vyaparis, or indigenous merchants, Kusavans or potters, Vannathamars or washermen of the non-polluting castes, and Chaliyas or weavers of silk cloths (Dames, 1921, Vol. 2:56–79).

### Polluting Castes

Ranking below the Nayars were the numerous polluting castes, which in relation to one another could be grouped into higher and lower. The higher polluting castes comprised the Tiyyas, the Kanisans, artisans, fishermen, and some of the service castes such as barbers and washermen, and probably the first eight castes of Barbosa's description.

Of the various higher polluting castes the Tiyyas constituted the single largest category, accounting for the bulk of all the polluting castes. Because of their low ritual status they suffered a lot of social disabilities, including untouchability and unapproachability (see, Rao, 1979:24). Their traditional occupation was planting and tapping of the coconut tree, though they also had a tradition of military service. But, as mostly share-cropping under-tenants of the Nambutiris, royal lineages, and the Nayars, and as agricultural labourers their economic position was also one of subservience.

Of the remaining higher polluting castes, Barbosa described the Mannans (also Vannans) as 'washermen for the common people, and makers of sleeping mats,' and 'slaves of the King and of the Nayres'; the Kanisans as great astrologers whose trade was to make bucklers and umbrellas; the Asaris as masons, carpenters, smiths, and metal workers, all of a common descent but each a separate caste; the Mukaver as sailors, fishers, and slaves of the Nayars; the Mukkuvans as fishermen; the Vettuvans as salt-makers, rice-growers and slaves of the kings and the Nayars; and the Panans (also Malayans) as great sorcerers, archers, and hunters (Dames, 1921, Vol. 2:61–67).

The Pulayas (also Cherumas) and Parayas who accounted for the bulk of the lower polluting castes were agrestic serfs of the Nambutiris and the Nayars.

Of the immigrants of Barbosa's description only the Mappilas are an important group in relation to the subject of this study. The emergence of this group has to be understood in the context of the commercial activities of the Arabs in Malabar. Starting from the thirteenth century A.D. most of these activities were concentrated in the kingdoms of Calicut in south Malabar and Cannanore in north Malabar. In the former, in addition to their commercial monopoly the Arabs were allowed to spread their religion. Conversion of the natives to Islam and inter-marriages led to the emergence of the Mappilas as a group of Malabar Muslims. They became most numerous in the Ernad and Valluvanad taluks in south Malabar.[5]

[5] Barbosa's description of the Mappilas reveals them as very numerous and prosperous: 'And they are so many and so rooted in the soil throughout Malabar that it seems to me they are a fifth part of its people spread over all

In the nineteenth century, following the large-scale conversion of freed slaves as well as some of the fisherfolk, most of the Mappilas along the Malabar coast were either fishermen or traders, while in the interior many had become tenants and agricultural labourers. As a group their place in the land-based social order was also one of socio-economic subservience to the Nambutiris and the Nayars.[6]

### *Relative Size*

The percentages of the different castes in the total population of the nine Malabar taluks, as tabulated from the 1871 Census of the Madras Presidency, give a general idea of the relative size of the principal social groups of Malabar in the pre-British times as well. Of the various groups mentioned in Table 2.2, the Brahmins include the Nambutiris and Pattars. Of these, the former accounted for 39 per cent and the latter for the rest of the Brahmin population. Logan mentioned that the former consisted of 1,017 families distributed as 277, 289, 120, 152, 70, 30, and 79 in the Valluvanad, Ponnani, Ernad, Calicut, Kurumbranad, Kottayam, and Chirakkal taluks respectively, and that the Palghat and Wynad taluks did not have any Nambutiri family at all (see, Government of Madras, 1882, Vol. 2:11). The Kshatriyas, also reported as warriors in the Census, seem to represent the royal lineages. The high-caste Nayars call for no explanation. The low-caste Nayars comprise the Census categories of Idaiyar (Shepherds), Kammalan (Artisans), Kanakkan (Writers), Kaikolan (Weavers), Vannian (Labourers), and Kusavan (Potters). While the Tiyyas also call for no explanation, the other higher polluting castes comprise the Census categories of Sembadavan (Fishermen), Ambattan (Barbers), and Vannan (Washermen). The lower polluting castes are the Cherumas and the Pariahs. The Muslims are mostly (92 per cent) the

---

its kingdoms and provinces. They are rich, and live well; they hold all the sea trade and navigation in such sort that if the King of Portugal had not discovered India, Malabar would already have been in the hands of the Moors, and would have had a Moorish King' (Dames, 1921, Vol. 2:74).

[6] For the emergence of the Mappilas as a group of Malabar Muslims with a separate social identity, see Dale, 1980:24–32; Houtart and Lemercinier, 1978:28.

**Table 2.2** ***Population Percentage by Caste / Community, Malabar, 1871***

| *Caste/Community* | *South Malabar* | | | | | *North Malabar* | | | | *Total* |
|---|---|---|---|---|---|---|---|---|---|---|
| | *Palghat* | *Vallu-vanad* | *Ponnani* | *Ernad* | *Cali-cut* | *Wynad* | *Kurum-branad* | *Kotta-yam* | *Chirak-kal* | |
| Brahmins | 6.6 | 1.5 | 1.1 | 0.5 | 1.4 | 0.7 | 0.7 | 0.9 | 1.7 | 1.9 |
| Kshatriyas | 0.1 | 0.0 | 0.0 | 0.0 | 0.1 | 0.1 | 0.0 | 0.0 | 0.0 | 0.0 |
| High-caste Nayars | 13.4 | 19.3 | 12.7 | 7.9 | 18.9 | 8.3 | 19.0 | 14.5 | 15.8 | 14.5 |
| Low-caste Nayars | 13.8 | 11.7 | 5.1 | 5.1 | 6.4 | 8.2 | 10.3 | 6.5 | 13.2 | 9.1 |
| Tiyyas | 32.4 | 15.3 | 22.5 | 12.1 | 27.3 | 3.3 | 30.5 | 41.2 | 26.6 | 23.5 |
| Other Higher Polluting Castes | 2.7 | 3.5 | 3.6 | 2.4 | 3.5 | 1.0 | 2.5 | 3.1 | 2.9 | 2.9 |
| Lower Polluting Castes | 12.7 | 18.9 | 13.8 | 18.2 | 13.7 | 20.2 | 8.5 | 1.1 | 9.1 | 13.3 |
| Muslims | 9.2 | 25.8 | 39.5 | 49.4 | 23.3 | 6.2 | 23.8 | 23.4 | 20.9 | 26.5 |
| Christians | 0.0 | 0.0 | 0.0 | 0.1 | 1.2 | 0.6 | 0.0 | 0.6 | 1.0 | 0.3 |
| Others | 9.0 | 4.0 | 1.6 | 4.3 | 4.2 | 51.4 | 4.6 | 8.6 | 8.8 | 8.0 |
| Total | 100.0 | 100.0 | 100.0 | 100.0 | 100.0 | 100.0 | 100.0 | 100.0 | 100.0 | 100.0 |

**Source:** Tabulated from the *Census of Madras Presidency, 1871*: Statement of Population of 1871 in each village of the Malabar District, Government of Madras, 1874.

Mappilas. The residual category of 'Others' consists of the unspecified Census category of 'other good castes,' the Satanis or mixed castes, and another unspecified Census category of 'Others' comprising the tribals. The last category explains the very high percentage of 'Others' in the Wynad taluk.

Although the Muslims, Tiyyas, Nayars, and the lower polluting castes together accounted for about four-fifths of the total population (27 per cent, 24 per cent, 15 per cent and 13 per cent respectively), in a society founded and functioning on the principles of social closure what really mattered was not so much the numerical preponderance of any group as its status in the land system reinforced by its corresponding status in the caste system.

Hard data are not available on this nexus between land status and caste status. But the data collected by Logan in 1881–82 on principal *janmis*, that is *janmis* owning at least 100 pieces of land in any one *amsam* or village (as tabulated in Table 2.3), give a general idea. Considering that the changes brought about by the British were conceptual and juridical, strengthening the position of the *janmis* and weakening that of the *kanakkar* and *verumpattakkar*, and not in the nature of large-scale redistribution of the principal land rights among the different social groups, a juxtaposed reading of these data and the population data given in Table 2.2 shows that this nexus was very close: the Brahmins were only about 2 per cent in the population, but nearly half (46 per cent) of the principal *janmis*, with estates lying chiefly in the southern taluks of Valluvanad, Ponnani, and Ernad; the royal lineages were an insignificant proportion in the population, but one in every fourteen principal *janmis*; and the high-caste Nayars were about one-seventh of the population, but two-fifths (41 per cent) of the principal *janmis*, fairly evenly distributed over the whole district. Stated differently, these three high-castes together were only about one-sixth of the total population, but accounted for nine out of every ten principal *janmis*. The remaining groups, mainly the lower castes and the Mappilas accounted for more than four-fifths of the population, but only less than one-sixteenth of the principal *janmis*.

Though only 7 per cent of the principal *janmis* were from the royal lineages, the importance of this group lies in the

**Table 2.3** ***Principal Janmis by Caste, Malabar, 1881–82 (Janmis in Number and Percentage)***

| *Taluk* | *Brahmins* | *Rajas* | *Nayars* | *Tiyyas* | *Other Hindus* | *Mappilas* | *Total* |
|---|---|---|---|---|---|---|---|
| *South Malabar* | | | | | | | |
| Palghat | 32 | 11 | 52 | — | 1 | — | 96 |
| | 33.3 | 11.5 | 54.2 | | 1.0 | | 100.0 |
| Valluvanad | 111 | 14 | 54 | — | — | — | 179 |
| | 62.0 | 7.8 | 30.2 | | | | 100.0 |
| Ponnani | 142 | 9 | 54 | 4 | — | 10 | 219 |
| | 64.8 | 4.1 | 24.7 | 1.8 | | 4.6 | 100.0 |
| Ernad | 62 | 6 | 43 | — | — | 2 | 113 |
| | 54.9 | 5.3 | 38.0 | | | 1.8 | 100.0 |
| Calicut | 11 | 10 | 34 | 1 | — | — | 56 |
| | 19.6 | 17.9 | 60.7 | 1.8 | | | 100.0 |
| *North Malabar* | | | | | | | |
| Wynad | — | — | 23 | — | 3 | — | 26 |
| | | | 88.5 | | 11.5 | | 100.0 |
| Kurum- | 7 | 6 | 44 | 1 | — | 8 | 66 |
| branad | 10.6 | 9.1 | 66.7 | 1.5 | | 12.1 | 100.0 |
| Kottayam | — | 2 | 18 | 2 | — | 10 | 32 |
| | | 6.3 | 56.2 | 6.3 | | 31.2 | 100.0 |
| Chirakkal | 14 | 3 | 17 | — | — | 7 | 41 |
| | 34.1 | 7.3 | 41.5 | | | 17.1 | 100.0 |
| Total | 379 | 61 | 339 | 8 | 4 | 37 | 828 |
| | 45.8 | 7.4 | 40.9 | 1.0 | 0.5 | 4.4 | 100.0 |

**Note:** Dash indicates nil. Of the Brahmins 370 are Malabar Brahmins (Nambutiris), 7 are Canara Brahmins (Embrantiris, in Chirakkal taluk), and 2 are East Coast Brahmins (Pattars, in Palghat taluk).

**Source:** Tabulated from Government of Madras, 1882, Vol. 1:lvi.

fact that these *janmis* held the largest estates. The average number of *amsams* in which they held at least 100 pieces of land was 4.9, as against 1.3 among the Nambutiris, and 1.4 among the Nayars (Government of Madras, 1882, Vol. 1:lvi).

Apart from these principal *janmis*, there were as many as 23,635 minor *janmis* (holding less than 100 pieces of land in any one *amsam*) in the district (Government of Madras, 1882, Vol. 1:lvii). Though Logan did not give the caste-wise break-up of these *janmis*, presumably most of them were also drawn from the Brahmins, royal lineages, and the high-caste Nayars.

The close nexus between the land status and caste status found in the distribution of principal *janmis* is also found in that of principal *janmam* institutions, that is, institutions with not less than 100 pieces of *janmam* land in any one *amsam*. As may be seen from Table 2.4, virtually all (96 per cent) of these institutions were also of the upper castes. Logan's observation that in north Malabar these institutions were rare, whereas their number was the largest in south Malabar where the Nambutiris settled most extensively (see, Government of Madras, 1882, Vol. 1:lviii), clearly indicates that these institutions were mostly of the Nambutiris.

Logan did not give the caste-wise distribution of the remaining land rights. However, his description of the land tenures shows that *kanam*, the second important land right, was held mostly by the high-caste Nayars, and that the lower castes, that is, the low-caste Nayars and the polluting castes, and the Mappilas held either only inferior land rights or no right at all. Thus, *verumpattam* was mostly held by the Tiyyas and the Mappilas in south Malabar and by the Tiyyas and the low-caste Nayars in north Malabar, and *cherujanmam* was mostly held by the artisan and service sections of the Nayars and the higher polluting castes in both south and north Malabar. The lower polluting castes mostly had no land right other than to work as agricultural labourers.

As already seen, the land system with these hierarchical land rights was highly iniquitous, oppressive, and exploitative. As will be evident from the following chapter, it was the further accentuation of this character of the land system since the second half of the eighteenth century which necessitated the long series of peasant struggles and land reforms in Malabar.

**Table 2.4** ***Religious and Charitable Institutions as Principal Janmis, Malabar, 1881–82 (Institutions in Number and Percentage)***

| *Taluk* | *Religious Institutions, High Castes* | *Religious Institutions, Muslims* | *Charitable Institutions, Hindus* | *Total* |
|---|---|---|---|---|
| *South Malabar* | | | | |
| Palghat | 43 | — | 3 | 46 |
| | 93.5 | | 6.5 | 100.0 |
| Valluvanad | 61 | — | 1 | 62 |
| | 98.4 | | 1.6 | 100.0 |
| Ponnani | 64 | — | 2 | 66 |
| | 97.0 | | 3.0 | 100.0 |
| Ernad | 29 | 1 | 3 | 33 |
| | 87.9 | 3.0 | 9.1 | 100.0 |
| Calicut | 6 | — | — | 6 |
| | 100.0 | — | — | 100.0 |
| *North Malabar* | | | | |
| Wynad | 11 | — | — | 11 |
| | 100.0 | | | 100.0 |
| Kurumbranad | 9 | — | — | 9 |
| | 100.0 | — | — | 100.0 |
| Kottayam | 1 | — | — | 1 |
| | 100.0 | — | — | 100.0 |
| Chirakkal | 16 | — | — | 16 |
| | 100.0 | — | — | 100.0 |
| Total | 240 | 1 | 9 | 250 |
| | 96.0 | 0.4 | 3.6 | 100.0 |

**Note:** Dash indicates nil.
**Source:** Tabulated from Government of Madras, 1882, Vol. 1:lix.

# 3

## Peasant Struggles and Land Reforms, 1836–1956

In discussing the long history of peasant struggles and land reforms in Malabar, it is necessary at the outset to clarify that the term peasantry has a very wide connotation when it is applied to Malabar in the past. Since virtually all the principal social groups participated in the process of agricultural production and subsisted on their customary shares of the produce of the soil, all of them (except the Nambutiris, royal lineages, and sections of the Nayars) qualify to be termed as peasants. This large class of peasantry constituted a hierarchy with Nayars as *kanakkar* at the top, Tiyyas and Mappilas as *verumpattakkar* in the middle, artisan and service castes as *cherujanmakkar* below them, and the lower polluting castes as agricultural labourers at the bottom. As will be evident shortly, it was the infringement of the customary land rights of these groups first by the rulers of Mysore and then by the British which caused the numerous peasant struggles.

### Mappila Outbreaks and Land Reforms: 1836–1921

In the first phase, these struggles were initiated and led by the Mappilas. There were some special reasons for this. Dale has argued that until the Portuguese discovery of Calicut in the year 1498 relations between the Mappilas and the Nambutiris and Nayars were characterised by 'limited contact and self-interested toleration' and that the Portuguese intrusion for monopolising the spice trade resulting in bitter struggles with the local Mappila traders disrupted this situation by precipitating major changes. Important among these changes were the creation of a militant tradition among the Mappilas, intensification of their awareness of their own identity which increased their sense of separateness from the rest of Malabar society, and lessening of their acceptance of, and respect for, traditional institutions and authority (Dale, 1980:32–65).

The occupation of Malabar by the Mysore rulers, Haidar Ali and Tipu Sultan, in the late eighteenth century further embittered the relations between the Mappilas and the caste-Hindus. Many Mappilas succeeded in shifting to their *janmis* the entire burden of the new land revenue assessment introduced in this period, especially in areas which were directly under the Mysore rule where many of the *janmis* were afraid of appearing at the revenue offices. More important than this was their possession of land when several of the Nambutiris and Nayars fled Malabar to escape religious persecution and forced conversion.[1] This possession, however, lasted only a few years, for when Malabar came under the British towards the close of the eighteenth century, the *janmis* rushed back and retrieved their lost lands with the help of the British government and its law courts (see Logan, 1951b:209–10). But the euphoria which the Mysore rule created among them and the new sense of freedom which they tasted lingered on and lent them the will to fight the system which they had until then passively accepted. Thus, if the Portuguese intrusion produced a militant tradition among the Mappilas, the invasions of the Mysore rulers prepared the ground for an outlet to this tradition through widespread agrarian discontent.

[1] For the impact of Mysore rule on Mappila-Hindu relations and Malabar society, see Dale 1980:79–91; Kareem, 1973; Kunju, 1975; Miller, 1976.

### British Land Policy

This discontent was deepened further by the British land policy of conferring absolute ownership rights on the *janmis*. Of this policy and its evil effects Logan wrote:

> The Bombay authorities first, and the Madras authorities afterwards, recognised the *Janmi* as absolute owner of his holding, and therefore free to take as big a share of the produce of the soil as he could screw out of the classes beneath him . . . . All customs limiting his freedom to deal as he liked with what was considered to be his own property in the European sense were construed with great strictness, and invariably gave way before express contract. Hence harder and harder terms were imposed by the *Janmi* on the *Kanakkaran* as time went on, and it was more and more clearly seen that the *Janmi* was no longer dependent as formerly on his body of *Kanakkar* for security in his rank and position. The British Courts backed up by Police and Magistrates and troops and big guns made the *Janmi*'s independence complete. The hard terms thus imposed on the *Kanakkaran* had, of course, the effect of hardening the terms imposed by the *Kanakkaran* on those below him, the *Verumpattakkar*. The one-third of the net produce to which the *Verumpattakkaran* was customarily entitled, was more and more encroached upon as the terms imposed on the *Kanakkaran* became harder and harder. (Government of Madras, 1882, Vol. 1:xvii, xxxi–ii)

For various reasons the crisis did not arise at once. But, as Logan wrote, it became inevitable 'as soon as the old generation with its inherited ideas of protection and trusts had died down, and as soon as prices of produce began to rise and hold out tempting baits to cupidity' (Government of Madras, 1882, Vol. 1:xvii). The Hindu tenants were passive in their suffering because of their submission to, and respect for, traditional authority. But the Mappilas who had inherited a militant tradition and in whom the Mysore rule had instilled a new sense of freedom and hope could not tolerate it. The result was a long series of protests by them which started as early as 1836 and lasted nearly a century. These protests often involved the murder of Hindu *janmis* which the disgruntled Mappilas regarded as a meritorious act. Known in British administrative parlance as Mappila outbreaks or outrages, these protests have been viewed differently as conflict on the Islamic frontier sparked by the Mappilas' 'intensely felt awareness of themselves as members of a distinct religious community' (Dale, 1980:5), as religious fanaticism of the Mappilas (Mazumdar, 1973; Ravindran, 1973); and as expressions of agrarian discontent (Houtart and Lemercinier, 1978; Namboodiripad, 1943:1–2; Panikkar, 1979:611; Wood, 1974). The manner in which the British authorities reacted to these protests, as will be discussed shortly, lends credence to the last view.

'Mappila unrest,' wrote Innes, 'reached its high watermark between the years 1836 and 1853. In this period of eighteen years no fewer than twenty-two outbreaks took place besides numerous abortive risings and conspiracies' (Innes, 1951:78). And, forced by a violent outbreak in 1852 in the relatively calm north Malabar, in February that year the government appointed T.L. Strange, a judge of the Sadar Adalat with long experience in Malabar, as the first Special Commissioner to inquire into these outbreaks.

### *Strange Commission*

In submitting his report in September 1852, Strange enumerated the Mappila outbreaks that occurred between 1836 and 1852. In its conclusion and in a supplemental letter, he made some important observations and suggestions on the need for regulating agrarian relations in Malabar. Stating that the Mappilas complained of Hindu *janmis* evicting them at shorter intervals than what was customarily prescribed, he stressed the need to adhere to the customary tenures which he felt had been conceived in 'much wisdom, foresight and fairness' but not always enforced by the British Courts. For avoiding potentially explosive agrarian disputes, he made two specific suggestions: 'the registration of deeds to prevent false litigation, and the resolution of tenure conflicts through the executive action of the Collector, rather than through the

Courts'. As Mappilas were widely engaged in hill-land cultivation and as under the prevalent method of yearly hill lands assessment they were subjected to extortion under threat of overassessment and constantly recurring contacts with Hindu officials, he also urged the introduction of permanent assessment of the hill lands (Dale, 1980:162).

Though Strange was 'specifically charged to ascertain the causes of the outbreaks and to make a special enquiry into the question of agrarian disputes,' it was with the caveat that the 'grand object of his appointment was "to secure to the Nair and Brahmin population the most ample protection and safety possible against the effects of Moplah fanaticism" ' (Dale, 1980:158). This made his observations and suggestions on agrarian relations only peripheral to the central theme of his main report. In it he 'rejected the idea that the outbreaks were due to agrarian disputes or more particularly to oppressive landlords'. Instead, he attributed them to the religious fanaticism of the Mappilas, and accordingly recommended severe repressive measures, such as disarming of the Mappila population, deportation of the insurgents and escheat of their property, fining of the Mappila population of the areas involved in the outbreaks and the creation of a special police force to be stationed in the centre of the disturbed areas (Dale, 1980:159–61).

Though the government thought that these measures were necessary only temporarily and implemented them with great alacrity to the total exclusion of his suggestions on agrarian relations, they were quite as much desired even after about three decades. For, in October 1880 the government received an anonymous petition purporting to come from 'certain Mussulmans, Nayars, Tiyyans, and men of other castes inhabiting Malabar'. It contained the oft-repeated complaint that the tenants were unjustly treated by the *janmis*, that the native Munsiffs sided with the *janmis* and that there was danger of fresh outbreaks. While similar petitions were presented to Strange when he was appointed Special Commissioner, the present petition was very emphatic in attributing the past outbreaks to the cruel oppression and exploitation of tenants by *janmis*, in warning of the occurrence of a 'terrible outbreak' and in urging the appointment of a Special

Commissioner to inquire into agrarian discontent.[2] Wigram, officiating district judge of south Malabar to whom the government forwarded this petition confidentially, was also very emphatic in attributing the outbreaks to agrarian discontent. Among other things he observed that Strange gave far too little weight to agrarian discontent as a cause of the outbreaks, that his suggestions were rather aimed at giving the *janmi* a cheaper remedy for recovering his rent, that agrarian discontent which still existed may at any time break out in isolated instances of fanaticism, and that the sympathy of the Mappila population was invariably on the side of the fanatics. Referring to the Travancore Proclamation issued in the time of Sir T. Madhava Rao in 1867 reaffirming the principles laid down in 1829 on the subject of the relationship between the *janmis* and tenants (viz. tenants who paid their rent and customary fees for renewal could not be ejected), he further observed that had a similar course been followed in Malabar, it would have checked the rapacity of the *janmi* and encouraged the tenant to pay his rent regularly and to live on terms of amity with his *janmi*.

The other two officials whose views the government sought confidentially—Logan, District Magistrate and Collector of Malabar, and Macgregor, British Resident in Travancore-Cochin and former Collector of Malabar—also reported that the essential nature of the Mappila outbreaks was agrarian, and that in some taluks there still existed an amount of serious agrarian discontent, which if not inquired into and promptly and effectively remedied may at any time assume proportions fatal to the peace of that part of the country.[3]

### Logan Commission

Following these reports, in February 1881 the government appointed Logan as the second Commissioner, specially to inquire into and report upon 'the general question of the tenure of land and tenant right in Malabar, and the alleged insufficiency of compensation offered by the landlords and

[2] The full text of the petition is appended to this chapter.

[3] For the correspondence on the petition, see Government of Madras, 1885.

awarded for land improvements made by tenants' (Government of Madras, 1882, Vol. 1:iii).

For the elaborate, extensive, and first ever systematic investigation into the land question in Malabar that was to follow for the next one and a half years, Logan used a wide variety of sources. Through a confidential circular issued to all officers in the Civil Judicial Establishment and to the heads of villages, he collected information on the customary rates given for tenants' improvements in the several localities, and the rates adopted in executing decrees of Court, which he found were not always those prescribed by custom. His extensive tour of the district for personal intercourse with the people in the different localities enabled him to obtain a full statement of their grievances through as many as 2,200 petitions submitted by 4,021 persons.[4] He examined the registers of all the Civil Courts in the district from 1862 to 1880. Finally, he compiled a 'statistical register' covering various details of land tenures, prices, crimes, and so on (Government of Madras, 1882, Vol. 1:iii–v).

[4] The petitions show that it was from the chief Mappila taluks that Logan obtained the full expression of the grievances:

| *Taluk* | *Petitions* | *Petitioners* | | | | | | | |
|---|---|---|---|---|---|---|---|---|---|
| | | *Bs* | *Ns* | *Ts* | *Cs* | *Ms* | *Os* | *Total* |
| Palghat | 169 | 25 | 32 | 141 | — | 9 | — | 207 |
| Valluvanad | 432 | 7 | 379 | 30 | — | 479 | — | 895 |
| Ponnani | 857 | 1 | 96 | 137 | 1 | 1,353 | 82 | 1,670 |
| Ernad | 418 | 6 | 16 | 23 | — | 767 | — | 812 |
| Calicut | 3 | — | — | 1 | — | 2 | — | 3 |
| Wynad | — | — | — | — | — | — | — | — |
| Kurumbranad | 258 | 13 | 19 | 150 | — | 110 | — | 292 |
| Kottayam | 28 | — | 1 | 19 | — | 8 | — | 28 |
| Chirakkal | 3 | — | 19 | — | — | 6 | — | 25 |
| Total | 2,168 | 52 | 562 | 501 | 1 | 2,734 | 82 | 3,932 |

**Note:** Dash indicates nil. Bs: Brahmins; Ns: Nayars; Ts: Tiyyas; Cs: Cherumas; Ms: Mappilas; Os: Others. Table excludes 30 anonymous petitions and two petitions from Cochin by 89 persons.

**Source:** Government of Madras, 1882, Vol. 1:ix.

In his report, prepared on the basis of the vast amount of rich data so collected, and submitted in June 1882, Logan indicted the British authorities for jeopardising the customary land tenures in Malabar. His most incisive and controversial statement was perhaps in regard to the British policy of conferring absolute ownership right on the *janmis*:

> [The] British authorities and British Courts have never properly realised the true original state of the relations subsisting between *Janmis* and the other classes interested in the soil. The *Janmi's* interest in the produce of the land was a distinctly limited interest, as were also those of the other classes. Their interests, regulated by ancient custom, should have been clearly ascertained and precisely defined before wholesale freedom of contract was let loose over the land . . . . The idea . . . that the *Janmi* is absolute proprietor of the soil, and as such entitled to take as big a share of the produce as he can command, is founded on error, and has been brought about by engrafting on the customary *Malayali* law usages, which spring from European ideas of property having a totally distinct origin and history. (Government of Madras, 1882, Vol. 1:xviii, xxvii)

Logan also dwelt at length on the British misconception of *kanam* as a lease or a mortgage or a mortgage-lease liable to renewal at the end of every 12 years, and *verumpattam* as a tenure from year to year, unless the lease specifically provided otherwise (see, Government of Madras, 1882, Vol. 1:xxvii–xxxii). His observations on the overall effects of these British aberrations on the principal agrarian classes may be summarised in his own words:

> The *Janmis* have profited too much, the other classes too little, by the British rule of personal freedom to follow one's legitimate inclinations . . . . The *Janmi* if content to gather in his rents annually instead of in lump sums every 12 years, would be the richer by the extravagant interest which the *Kanakkaran* has now to pay to the money-lender. The *Kanakkaran* starting with a load of debt does not benefit one pie until the lump sum paid for renewal

> fees has been cleared off with extortionate interest thereon. Finally, the actual cultivator, if there is such a person below the *Kanakkaran*, is saddled with an extreme rack-rent, in order to help the *Kanakkaran* to pay off the money-lender's claims. A system could hardly be devised better adapted in every way for ruining the agricultural class. (Government of Madras, 1882, Vol. 1:xviii, xxix–xxx)

For understanding the nature of the existing land tenures Logan collected data from as many as 14,034 different pieces of land. These lands accounted for about 96 per cent of the 14,567 pieces of land owned by 110 typical *janmis* of all classes—good, bad, and indifferent—and were held by 7,994 actual cultivators, either direct from *janmis* or under intermediaries (Table 3.1). Logan classified the tenures of these lands into five broad categories:

1. Permanent tenures with or without rent;
2. (*a*) tenures with advances secured on the land, but without rent, not liable to renewal fees at fixed intervals (*otti*);
   (*b*) tenures with advances secured on the land, but without rent, liable to renewal fees at fixed intervals (quasi-*otti*);
3. (*a*) tenures with advances secured on the land, and with rent, liable to renewal fees at fixed intervals (*kanam*);
   (*b*) tenures with advances secured on the land, and with rent, not liable to renewal fees at fixed intervals (quasi-*kanam*);
4. mortgages with or without rent for definite or indefinite periods (*panayam*);
5. (*a*) leases for 12 years or more (*verumpattam*);
   (*b*) leases for more than one year and less than 12 years (*verumpattam*); and
   (*c*) leases for one year or at will (*verumpattam*). (Government of Madras 1882, Vol. 1:xxxiii)

The distribution of the lands into these tenures as revealed by his data may be summarised in the following way.

**Table 3.1** ***Lands by Type of Land Right, Malabar, 1881–82***

| *Taluk* | *No. of Janmis* | *No. of Inter-mediaries under Janmis* | *No. of Actual Cultivators Under* | | *Lands Leased Out by Janmis to* | | *Lands Retained by Janmis* | *Total Lands Owned by Janmis* |
|---|---|---|---|---|---|---|---|---|
| | | | *Janmis* | *Inter-mediaries* | *Inter-mediaries* | *Actual Cultivators* | | |
| *South Malabar* | | | | | | | | |
| Palghat | 12 | 975 | 562 | 590 | 922 | 920 | 6 | 1,848 |
| | | | | | 49.9 | 49.8 | 0.3 | 100.0 |
| Valluvanad | 12 | 511 | 603 | 309 | 476 | 1,108 | 27 | 1,611 |
| | | | | | 29.5 | 68.8 | 1.7 | 100.0 |
| Ponnani | 16 | 1,270 | 719 | 420 | 1,046 | 1,189 | 271 | 2,506 |
| | | | | | 41.7 | 47.5 | 10.8 | 100.0 |
| Ernad | 15 | 414 | 577 | 256 | 386 | 898 | 57 | 1,341 |
| | | | | | 28.8 | 67.0 | 4.2 | 100.0 |
| Calicut | 6 | 106 | 291 | 74 | 102 | 589 | 23 | 714 |
| | | | | | 14.3 | 82.5 | 3.2 | 100.0 |

Table 3.1 *Contd.*

| *Taluk* | *No. of Janmis* | *No. of Inter-mediaries under Janmis* | *No. of Actual Cultivators Under* | | *Lands Leased Out by Janmis to* | | *Lands Retained by Janmis* | *Total Lands Owned by Janmis* |
|---|---|---|---|---|---|---|---|---|
| | | | *Janmis* | *Inter-mediaries* | *Inter-mediaries* | *Actual Cultivators* | | |
| *North Malabar* | | | | | | | | |
| Wynad | 12 | 37 | 503 | 27 | 37 | 897 | Nil | 934 |
| | | | | | 4.0 | 96.0 | | 100.0 |
| Kurumbranad | 11 | 342 | 599 | 258 | 320 | 1,035 | 36 | 1,391 |
| | | | | | 23.0 | 74.4 | 2.6 | 100.0 |
| Kottayam | 12 | 383 | 441 | 181 | 365 | 676 | Nil | 1,041 |
| | | | | | 35.1 | 64.9 | | 100.0 |
| Chirakkal | 14 | 56 | 1,550 | 34 | 52 | 3,016 | 113 | 3,181 |
| | | | | | 1.6 | 94.8 | 3.6 | 100.0 |
| Total | 110 | 4,094 | 5,845 | 2,149 | 3,706 | 10,328 | 533 | 14,567 |
| | | | | | 25.4 | 70.9 | 3.7 | 100.0 |

**Note:** All the lands leased in by intermediaries were leased out to the actual cultivators under them as shown in the Table.
**Source:** Tabulated from Government of Madras, 1882, Vol. 2:Appendix iv.

1. Of the 14,034 pieces of wet/dry and garden lands[5] leased in by actual cultivators, as many as 10,328 (74 per cent) were direct from *janmis* and only 3,706 (26 per cent) were from intermediaries (Table 3.1);
2. Of the wet/dry and garden lands leased in by actual cultivators direct from *janmis*, permanent tenures accounted for only an insignificant proportion (3 per cent) and *kanam* tenures only for about one-third (34 per cent), whereas *verumpattam* tenures accounted for nearly two-thirds (61 per cent) (Table 3.2);
3. Of the *verumpattam* tenures about four-fifths were on short durations of 12 years or less, with about half of them of less than one year or at will (Table 3.2);
4. Of the lands leased in by actual cultivators from intermediaries, permanent and *kanam* tenures accounted for very small proportions, whereas *verumpattam* tenures accounted for as high as three-fourths, with virtually all of them of the duration of less than one year or at will (Table 3.3);
5. Of the 3,706 pieces of wet/dry and garden lands leased in by actual cultivators from intermediaries, permanent tenures and *verumpattam* tenures accounted for only very small proportions, whereas *kanam* tenures accounted for the bulk (82 per cent) (Table 3.4);
6. In north Malabar *kanam* was almost extinct (except in the case of garden lands in the Kurumbranad taluk) as a means of leasing lands between *janmis* and actual cultivators, though it still survived there to a considerable extent as a means of leasing lands between *janmis* and intermediaries. In contrast to this, in south Malabar a large proportion of the lands held direct from *janmis* by actual cultivators and from *janmis* by intermediaries were still on *kanam* tenures (Tables 3.2 and 3.4);

[5] According to Logan's account, the wet/dry lands were the grain lands (paddy and other crops), or what may be called *nanja* (wet lands) and *punja* (dry lands). *Nanja* lands were almost exclusively used for paddy cultivation. The chief *punja* crops in Malabar were *punam, modan, palli manyayal*, and gingelly. *Punam* was chiefly cultivated on hills covered with jungle. It was originally raised only once in 12 years, mostly by the poor people. Other crops were cultivated in level waste *parambas* and sometimes in gardens. The chief garden crops were coconut, jackfruit, areca, and pepper in their order of importance. See Government of Madras, 1882, Vol. 2:387–96.

Table 3.2 *Holdings of Actual Cultivators under Janmis by Type of Tenure, Malabar, 1881–82 (Holdings in Number and Percentage)*

| *Taluk* | *Wet/Dry Land* | | | | | | | *Garden Land* | | | | | | |
|---|---|---|---|---|---|---|---|---|---|---|---|---|---|---|
| | *Kanam* | *Verumpattam* | | | *Permanent Tenures* | *Others* | *Total* | *Kanam* | *Verumpattam* | | | *Permanent Tenures* | *Others* | *Total* |
| | | *a* | *b* | *c* | | | | | *a* | *b* | *c* | | | |
| *South Malabar* | | | | | | | | | | | | | | |
| Palghat | 559 | 24 | — | 220 | 61 | 7 | 871 | — | — | — | 49 | — | — | 49 |
| | 64.2 | 2.7 | | 25.3 | 7.0 | 0.8 | 100.0 | | | | 100.0 | | | 100.0 |
| Valluvanad | 494 | — | — | 274 | 13 | 15 | 796 | 202 | — | — | 66 | 37 | 7 | 312 |
| | 62.1 | | | 34.4 | 1.6 | 1.9 | 100.0 | 64.7 | | | 21.2 | 11.9 | 2.2 | 100.0 |
| Ponnani | 257 | 24 | 107 | 167 | 14 | 47 | 616 | 425 | 13 | 51 | 29 | 38 | 17 | 573 |
| | 41.7 | 3.9 | 17.4 | 27.1 | 2.3 | 7.6 | 100.0 | 74.2 | 2.2 | 8.9 | 5.1 | 6.6 | 3.0 | 100.0 |
| Ernad | 244 | — | 69 | 71 | — | 8 | 392 | 420 | — | 14 | 62 | — | 10 | 506 |
| | 62.3 | | 17.6 | 18.1 | | 2.0 | 100.0 | 83.0 | | 2.8 | 12.2 | | 2.0 | 100.0 |
| Calicut | 207 | 47 | 3 | 109 | — | 1 | 367 | 149 | 42 | — | 31 | — | — | 222 |
| | 56.4 | 12.8 | 0.8 | 29.7 | | 0.3 | 100.0 | 67.1 | 18.9 | | 14.0 | | | 100.0 |

Table 3.2 *Contd.*

| *Taluk* | *Wet/Dry Land* | | | | | | | *Garden Land* | | | | | | |
|---|---|---|---|---|---|---|---|---|---|---|---|---|---|---|
| | *Kanam* | *Verumpattam* | | | *Permanent Tenures* | *Others* | *Total* | *Kanam* | *Verumpattam* | | | *Permanent Tenures* | *Others* | *Total* |
| | | *a* | *b* | *c* | | | | | *a* | *b* | *c* | | | |
| *North Malabar* | | | | | | | | | | | | | | |
| Wynad | 25 | 83 | 26 | 147 | 63 | 8 | 352 | 9 | 73 | 105 | 290 | 61 | 7 | 545 |
| | 7.1 | 23.6 | 7.4 | 41.7 | 17.9 | 2.3 | 100.0 | 1.6 | 13.4 | 19.3 | 53.2 | 11.2 | 1.3 | 100.0 |
| Kurumbranad | 127 | 112 | 95 | 35 | — | — | 369 | 370 | 203 | 40 | 52 | — | 1 | 666 |
| | 34.4 | 30.4 | 25.7 | 9.5 | | | 100.0 | 55.6 | 30.5 | 6.0 | 7.8 | | 0.1 | 100.0 |
| Kottayam | — | — | 421 | 6 | — | 18 | 445 | -- | 168 | 59 | 4 | — | — | 231 |
| | | | 94.6 | 1.4 | | 4.0 | 100.0 | | 72.7 | 25.5 | 1.7 | | | 100.0 |
| Chirakkal | 6 | 4 | 869 | 791 | 35 | 23 | 1,728 | 1 | 179 | 893 | 186 | 16 | 13 | 1,288 |
| | 0.3 | 0.2 | 50.3 | 45.8 | 2.0 | 1.3 | 100.0 | 0.1 | 13.9 | 69.3 | 14.4 | 1.2 | 1.0 | 100.0 |
| Total | 1,919 | 294 | 1,590 | 1,820 | 186 | 127 | 5,936 | 1,576 | 678 | 1,162 | 769 | 152 | 55 | 4,392 |
| | 32.3 | 4.9 | 26.8 | 30.7 | 3.1 | 2.1 | 100.0 | 35.9 | 15.4 | 26.5 | 17.5 | 3.5 | 1.3 | 100.0 |

**Note:** Dash indicates nil. Wet/dry *kanam* holdings include 17 quasi-*kanam* holdings; 'Others' are 27 *otti*, 15 quasi-*otti* and 85 *panayam* holdings in the case of wet/dry land; and 6 *otti*, 11 quasi-*otti* and 38 *panayam* holdings in the case of garden land.

**Source:** Tabulated from Government of Madras, 1882, Vol. 2, Appendix iv.

Table 3.3 *Holdings of Actual Cultivators under Intermediaries by Type of Tenure, Malabar, 1881–82 (Holdings in Number and Percentage)*

| *Taluk* | *Wet/Dry Land* | | | | | | | *Garden Land* | | | | | | |
|---|---|---|---|---|---|---|---|---|---|---|---|---|---|---|
| | *Kanam* | *Verumpattam* | | | *Permanent Tenures* | *Others* | *Total* | *Kanam* | *Verumpattam* | | | *Permanent Tenures* | *Others* | *Total* |
| | | *a* | *b* | *c* | | | | | *a* | *b* | *c* | | | |
| ***South Malabar*** | | | | | | | | | | | | | | |
| **Palghat** | 33 | 1 | — | 768 | 3 | 117 | 922 | — | — | — | — | — | — | — |
| | 3.6 | 0.1 | | 83.3 | 0.3 | 12.7 | 100.0 | | | | | | | |
| **Valluvanad** | 11 | 3 | — | 351 | — | 70 | 435 | 13 | — | — | 23 | — | 5 | 41 |
| | 2.5 | 0.7 | | 80.7 | | 16.1 | 100.0 | 31.7 | | | 56.1 | | 12.2 | 100.0 |
| **Ponnani** | 8 | — | 1 | 777 | — | 84 | 870 | 3 | — | 2 | 139 | — | 32 | 176 |
| | 0.9 | | 0.1 | 89.3 | | 9.7 | 100.0 | 1.7 | | 1.1 | 79.0 | | 18.2 | 100.0 |
| **Ernad** | 11 | — | 100 | 161 | — | 46 | 318 | 26 | — | 5 | 24 | — | 13 | 68 |
| | 3.5 | | 31.4 | 50.6 | | 14.5 | 100.0 | 38.2 | | 7.4 | 35.3 | | 19.1 | 100.0 |
| Calicut | 18 | 17 | 1 | 10 | — | 12 | 58 | 14 | 3 | — | 6 | 2 | 19 | 44 |
| | 31.0 | 29.3 | 1.7 | 17.2 | | 20.7 | 100.0 | 31.8 | 6.8 | | 13.6 | 4.5 | 43.2 | 100.0 |

Table 3.3 *Contd.*

| *Taluk* | *Wet/Dry Land* | | | | | | | *Garden Land* | | | | | | |
|---|---|---|---|---|---|---|---|---|---|---|---|---|---|---|
| | *Kanam* | *Verumpattam* | | | *Permanent Tenures* | *Others* | *Total* | *Kanam* | *Verumpattam* | | | *Permanent Tenures* | *Others* | *Total* |
| | | *a* | *b* | *c* | | | | | *a* | *b* | *c* | | | |
| ***North Malabar*** | | | | | | | | | | | | | | |
| **Wynad** | — | — | — | 20 | — | — | 20 | — | — | — | 17 | — | — | 17 |
| | | | | 100.0 | | | 100.0 | | | | 100.0 | | | 100.0 |
| **Kurumbranad** | 12 | 32 | 8 | 26 | — | — | 78 | 93 | 123 | 10 | 16 | — | — | 242 |
| | 15.4 | 41.0 | 10.3 | 33.3 | | | 100.0 | 38.4 | 50.8 | 4.1 | 6.6 | | | 100.0 |
| **Kottayam** | 80 | — | 63 | 11 | — | 19 | 173 | 121 | 65 | 1 | 3 | 2 | — | 192 |
| | 46.2 | | 36.4 | 6.4 | | 11.0 | 100.0 | 63.0 | 33.9 | 0.5 | 1.6 | 1.0 | | 100.0 |
| **Chirakkal** | — | — | 42 | 8 | — | — | 50 | — | 1 | — | — | — | 1 | 2 |
| | | | 84.0 | 16.0 | | | 100.0 | | 50.0 | | | | 50.0 | 100.0 |
| **Total** | 173 | 53 | 215 | 2,132 | 3 | 348 | 2,924 | 270 | 192 | 18 | 228 | 4 | 70 | 782 |
| | 5.9 | 1.8 | 7.4 | 72.9 | 0.1 | 11.9 | 100.0 | 34.5 | 24.5 | 2.3 | 29.2 | 0.5 | 9.0 | 100.0 |

**Note:** Dash indicates nil. 'Others' are 19 *otti* and 329 *panayam* holdings in the case of wet/dry land, and 70 *panayam* holdings in the case of garden land.
**Source:** Tabulated from Government of Madras, 1882, Vol. 2, Appendix iv.

Table 3.4 *Holdings of Intermediaries under Janmis by Type of Tenure, Malabar, 1881–82 (Holdings in Number and Percentage)*

| *Taluk* | *Wet/Dry Land* | | | | | | | *Garden Land* | | | | | | |
|---|---|---|---|---|---|---|---|---|---|---|---|---|---|---|
| | *Kanam* | *Verumpattam* | | | *Permanent Tenures* | *Others* | *Total* | *Kanam* | *Verumpattam* | | | *Permanent Tenures* | *Others* | *Total* |
| | | *a* | *b* | *c* | | | | | *a* | *b* | *c* | | | |
| ***South Malabar*** | | | | | | | | | | | | | | |
| **Palghat** | 784 | 24 | — | 12 | 75 | 27 | 922 | — | — | — | — | — | — | — |
| | 85.0 | 2.6 | | 1.3 | 8.1 | 2.9 | 100.0 | | | | | | | |
| **Valluvanad** | 427 | — | — | 1 | 4 | 3 | 435 | 40 | — | — | — | 1 | — | 41 |
| | 98.2 | | | 0.2 | 0.9 | 0.7 | 100.0 | 97.6 | | | | 2.4 | | 100.0 |
| **Ponnani** | 753 | 5 | — | 7 | — | 105 | 870 | 161 | 1 | — | 2 | — | 12 | 176 |
| | 86.5 | 0.6 | | 0.8 | | 12.1 | 100.0 | 91.5 | 0.6 | | 1.1 | | 6.8 | 100.0 |
| **Ernad** | 183 | — | 100 | 4 | — | 31 | 318 | 54 | — | 3 | — | — | 11 | 68 |
| | 57.5 | | 31.4 | 1.3 | | 9.8 | 100.0 | 79.4 | | 4.4 | | | 16.2 | 100.0 |
| **Calicut** | 55 | — | — | 3 | — | — | 58 | 41 | — | — | — | 3 | — | 44 |
| | 94.8 | | | 5.2 | | | 100.0 | 93.2 | | | | 6.8 | | 100.0 |

Table 3.4 *Contd.*

| *Taluk* | *Wet/Dry Land* | | | | | | | *Garden Land* | | | | | | |
|---|---|---|---|---|---|---|---|---|---|---|---|---|---|---|
| | *Kanam* | *Verumpattam* | | | *Permanent Tenures* | *Others* | *Total* | *Kanam* | *Verumpattam* | | | *Permanent Tenures* | *Others* | *Total* |
| | | *a* | *b* | *c* | | | | | *a* | *b* | *c* | | | |
| ***North Malabar*** | | | | | | | | | | | | | | |
| **Wynad** | 8 | — | — | — | — | 12 | 20 | 9 | — | — | — | 1 | 7 | 17 |
| | 40.0 | | | | | 60.0 | 100.0 | 52.9 | | | | 5.9 | 41.2 | 100.0 |
| Kurumbranad | 54 | 14 | 2 | — | — | 8 | 78 | 192 | 38 | — | — | — | 12 | 242 |
| | 69.2 | 17.9 | 2.6 | | | 10.3 | 100.0 | 79.3 | 15.7 | | | | 5.0 | 100.0 |
| **Kottayam** | 108 | 1 | 3 | 2 | — | 59 | 173 | 181 | 2 | 1 | 1 | 2 | 5 | 192 |
| | 62.4 | 0.6 | 1.7 | 1.2 | | 34.1 | 100.0 | 94.3 | 1.0 | 0.5 | 0.5 | 1.0 | 2.6 | 100.0 |
| **Chirakkal** | 4 | — | — | 5 | — | 41 | 50 | — | 1 | — | — | — | 1 | 2 |
| | 8.0 | | | 10.0 | | 82.0 | 100.0 | | 50.0 | | | | 50.0 | 100.0 |
| **Total** | 2,376 | 44 | 105 | 34 | 79 | 286 | 2.924 | 678 | 42 | 4 | 3 | 7 | 48 | 782 |
| | 81.2 | 1.5 | 3.6 | 1.2 | 2.7 | 9.8 | 100.0 | 86.7 | 5.4 | 0.5 | 0.4 | 0.8 | 6.1 | 100.0 |

**Note:** Dash indicates nil. *Kanam* holdings include 8 quasi-*kanam* holdings in the case of wet/dry land, and 70 quasi-*kanam* holdings in the case of garden land. 'Others' are 116 *otti*, 9 quasi-*otti* and 161 *panayam* holdings in the case of wet/dry land, and 15 *otti* and 33 *panayam* holdings in the case of garden land.

**Source:** Tabulated from Government of Madras, 1882, Vol. 2. Appendix xiv.

7. The preponderance of *verumpattam* short-tenures and the variations in the incidence of *kanam* between the north and south Malabar taluks indicate that in the district in general, and in north Malabar in particular, there was a tendency to abandon *kanam* tenures in favour of ordinary tenures, especially tenures from year to year or at will. This tendency was also dwelt upon as a specific grievance in many of the petitions received by Logan. It was also evident from the data collected by him on the length of possession of lands by the families of the actual cultivators.

Of the wet/dry and garden lands held by the actual cultivators from both *janmis* and intermediaries only 4 per cent were held from time immemorial; while about 37 per cent were held for over 30 years, and about 59 per cent were held for 30 years or less, with about 66 per cent of them for less than 12 years (Table 3.5). Of the wet/dry and garden lands held for over 30 years, nearly half (46 per cent) were on *kanam* tenures (Tables 3.6 and 3.7). These and the *kanam* intermediaries under *janmis* represented probably the nucleus of the old protector guild, which was slowly passing away. The lands they retained longest under the new regime were their gardens and house-sites. Many of them, too, as the petitioners represented to Logan, had been compelled to exchange their *kanam* tenures for *verumpattam* short-leases, and this too accounted for the large number of *verumpattam* leases which had been in possession for periods extending over 30 years.

Logan felt that the tendency to abandon *kanam* tenure in favour of *verumpattam* leases was just what might be expected to happen. For, the ordinary lease was much more profitable to the *janmi* because the profits which, under *kanam* tenure, went into the pockets of the money lenders, found their way under the ordinary lease into the pockets of the *janmis* though the latter had more difficulty in collecting their rents. He also felt that it would become more and more noticeable, especially in the case of grain lands, because the grain lands were much more easily rack-rented than the gardens.

From an analysis of the eviction and rent suits in all the

**Table 3.5** *Holdings of Actual Cultivators by Length of Possession, Malabar, 1881–82 (Holdings in Number and Percentage)*

| *Taluk* | *Wet/Dry Land* | | | | | *Garden Land* | | | | |
|---|---|---|---|---|---|---|---|---|---|---|
| | *Immemorial* | *Above 30 Years but not immemorial* | *12 to 30 Years* | *Under 12 Years* | *Total* | *Immemorial* | *Above 30 Years but not immemorial* | *12 to 30 Years* | *Under 12 Years* | *Total* |
| *South Malabar* | | | | | | | | | | |
| Palghat | — | 618 | 264 | 911 | 1,793 | — | 43 | 3 | 3 | 49 |
| | | 34.5 | 14.7 | 50.8 | 100.0 | | 87.8 | 6.1 | 6.1 | 100.0 |
| Valluvanad | 9 | 506 | 157 | 559 | 1,231 | 26 | 223 | 41 | 63 | 353 |
| | 0.7 | 41.1 | 12.8 | 45.4 | 100.0 | 7.4 | 63.2 | 11.6 | 17.8 | 100.0 |
| Ponnani | 107 | 186 | 295 | 895 | 1,483 | 52 | 222 | 185 | 290 | 749 |
| | 7.2 | 12.5 | 19.9 | 60.4 | 100.0 | 6.9 | 29.6 | 24.7 | 38.7 | 100.0 |
| Ernad | — | 221 | 93 | 334 | 648 | — | 264 | 152 | 145 | 561 |
| | | 34.1 | 14.4 | 51.5 | 100.0 | | 47.9 | 27.1 | 25.9 | 100.0 |
| Calicut | — | 186 | 98 | 141 | 425 | — | 118 | 59 | 89 | 266 |
| | | 43.8 | 23.1 | 33.2 | 100.0 | | 44.4 | 22.2 | 33.5 | 100.0 |

Table 3.5 *Contd.*

| *Taluk* | *Wet/Dry Land* | | | | | *Garden Land* | | | | |
|---|---|---|---|---|---|---|---|---|---|---|
| | *Immemorial* | *Above 30 Years but not immemorial* | *12 to 30 Years* | *Under 12 Years* | *Total* | *Immemorial* | *Above 30 Years but not immemorial* | *12 to 30 Years* | *Under 12 Years* | *Total* |
| *North Malabar* | | | | | | | | | | |
| Wynad | 138 | 105 | 35 | 94 | 372 | 109 | 62 | 109 | 282 | 562 |
| | 37.1 | 28.2 | 9.4 | 25.3 | 100.0 | 19.4 | 11.0 | 19.4 | 50.2 | 100.0 |
| Kurumbranad | — | 261 | 80 | 106 | 447 | — | 557 | 145 | 206 | 908 |
| | | 58.4 | 17.9 | 23.7 | 100.0 | | 61.3 | 16.0 | 22.7 | 100.0 |
| Kottayam | — | 326 | 131 | 161 | 618 | — | 188 | 92 | 143 | 423 |
| | | 52.7 | 21.2 | 26.1 | 100.0 | | 44.4 | 21.8 | 33.8 | 100.0 |
| Chirakkal | 63 | 609 | 487 | 616 | 1,775 | 16 | 546 | 366 | 362 | 1,290 |
| | 3.6 | 34.3 | 27.4 | 34.7 | 100.0 | 1.2 | 42.3 | 28.4 | 28.1 | 100.0 |
| Total | 317 | 3,018 | 1,640 | 3,817 | 8,792 | 203 | 2,223 | 1,152 | 1,583 | 5,161 |
| | 3.6 | 34.3 | 18.7 | 43.4 | 100.0 | 3.9 | 43.1 | 22.3 | 30.7 | 100.0 |

**Note:** Dash indicates nil. Length of possession was not reported in the case of 3 wet/dry lands in Ponnani, 62 wet/dry lands and 13 garden lands in Ernad, and 3 wet/dry lands in Chirakkal.

**Source:** Tabulated from Government of Madras, 1882, Vol. 2, Appendix iv.

Table 3.6 *Wet/Dry Land Holdings of Actual Cultivators by Length of Possession and Type of Tenure, Malabar, 1881–82 (Holdings in Number and Percentage)*

| *Taluk* | *Above 30 Years but not Immemorial* | | | | | | *12 to 30 Years* | | | | | | *Under 12 Years* | | | | | |
|---|---|---|---|---|---|---|---|---|---|---|---|---|---|---|---|---|---|---|
| | *Kanam* | *Verumpattam* | | | *Others* | *Total* | *Kanam* | *Verumpattam* | | | *Others* | *Total* | *Kanam* | *Verumpattam* | | | *Others* | *Total* |
| | | *a* | *b* | *c* | | | | *a* | *b* | *c* | | | | *a* | *b* | *c* | | |
| *South Malabar* | | | | | | | | | | | | | | | | | | |
| Palghat | 445 | 17 | — | 80 | 76 | 618 | 64 | 6 | — | 168 | 26 | 264 | 85 | 3 | — | 732 | 91 | 911 |
| | 72.0 | 2.8 | | 12.9 | 2.3 | 100.0 | 24.2 | 2.3 | | 63.6 | 9.8 | 100.0 | 9.3 | 0.3 | | 80.4 | 10.0 | 100.0 |
| Valluvanad | 386 | 2 | — | 102 | 16 | 506 | 53 | — | — | 90 | 14 | 157 | 55 | 1 | — | 444 | 59 | 559 |
| | 76.3 | 0.4 | | 20.1 | 3.2 | 100.0 | 33.8 | | | 57.3 | 8.9 | 100.0 | 9.8 | 0.2 | | 79.4 | 10.6 | 100.0 |
| Ponnani | 72 | — | 3 | 92 | 19 | 186 | 55 | — | 15 | 202 | 23 | 295 | 73 | 6 | 90 | 635 | 91 | 895 |
| | 38.7 | | 1.6 | 49.5 | 10.2 | 100.0 | 18.6 | | 5.1 | 68.5 | 7.8 | 100.0 | 8.2 | 0.7 | 10.1 | 70.9 | 10.2 | 100.0 |
| Ernad | 161 | — | 26 | 29 | 5 | 221 | 41 | — | 16 | 21 | 15 | 93 | 55 | — | 65 | 178 | 36 | 334 |
| | 72.8 | | 11.8 | 13.1 | 2.3 | 100.0 | 44.1 | | 17.2 | 22.6 | 16.1 | 100.0 | 16.5 | | 19.5 | 53.3 | 10.8 | 100.0 |
| Calicut | 126 | 34 | — | 25 | 1 | 186 | 43 | 17 | — | 36 | 2 | 98 | 56 | 13 | 4 | 58 | 10 | 141 |
| | 67.7 | 18.3 | | 13.4 | 0.5 | 100.0 | 43.9 | 17.3 | | 36.7 | 2.0 | 100.0 | 39.7 | 9.2 | 2.8 | 41.1 | 7.1 | 100.0 |

Table 3.6 *Contd.*

| *Taluk* | *Above 30 Years but not Immemorial* | | | | | | *12 to 30 Years* | | | | | | *Under 12 Years* | | | | | |
|---|---|---|---|---|---|---|---|---|---|---|---|---|---|---|---|---|---|---|
| | *Kanam* | *Verumpattam* | | | *Others* | *Total* | *Kanam* | *Verumpattam* | | | *Others* | *Total* | *Kanam* | *Verumpattam* | | | *Others* | *Total* |
| | | *a* | *b* | *c* | | | | *a* | *b* | *c* | | | | *a* | *b* | *c* | | |
| *North Malabar* | | | | | | | | | | | | | | | | | | |
| Wynad | 3 | 72 | 1 | 23 | 6 | 105 | 2 | 11 | — | 18 | 4 | 35 | 9 | — | 25 | 56 | 4 | 94 |
| | 2.9 | 68.6 | 0.9 | 21.9 | 5.7 | 100.0 | 5.7 | 31.4 | | 51.4 | 11.4 | 100.0 | 9.6 | | 26.6 | 59.6 | 4.2 | 100.0 |
| Kurumbranad | 92 | 96 | 59 | 14 | — | 261 | 19 | 21 | 22 | 18 | — | 80 | 27 | 26 | 24 | 29 | — | 106 |
| | 35.2 | 36.8 | 22.6 | 5.4 | | 100.0 | 23.8 | 26.2 | 27.5 | 22.5 | | 100.0 | 25.5 | 24.5 | 22.6 | 27.4 | | 100.0 |
| Kottayam | 63 | — | 223 | 9 | 31 | 326 | 12 | — | 113 | 3 | 3 | 131 | 5 | — | 148 | 5 | 3 | 161 |
| | 19.3 | | 68.4 | 2.8 | 9.5 | 100.0 | 9.2 | | 86.3 | 2.3 | 2.3 | 100.0 | 3.1 | | 91.9 | 3.1 | 1.9 | 100.0 |
| Chirakkal | 3 | 3 | 387 | 200 | 16 | 609 | 1 | — | 232 | 250 | 4 | 487 | 1 | 1 | 254 | 346 | 14 | 616 |
| | 0.5 | 0.5 | 63.5 | 32.8 | 2.6 | 100.0 | 0.2 | | 47.6 | 51.3 | 0.8 | 100.0 | 0.2 | 0.2 | 41.2 | 56.2 | 2.3 | 100.0 |
| Total | 1,351 | 224 | 699 | 574 | 170 | 3,018 | 290 | 55 | 398 | 806 | 91 | 1,640 | 366 | 50 | 610 | 2,483 | 308 | 3,817 |
| | 44.8 | 7.4 | 23.2 | 19.0 | 5.6 | 100.0 | 17.7 | 3.4 | 24.3 | 49.1 | 5.5 | 100.0 | 9.6 | 1.3 | 16.0 | 65.1 | 8.1 | 100.0 |

**Note:** Dash indicates nil. 'Others' includes holdings under permanent tenures.
**Source:** Tabulated from Government of Madras, 1882, Vol. 2, Appendix iv.

Table 3.7 *Garden Land Holdings of Actual Cultivators by Length of Possession and Type of Tenure, Malabar, 1881–82 (Holdings in Number and Percentage)*

| *Taluk* | *Above 30 Years but not Immemorial* | | | | | | *12 to 30 Years* | | | | | | *Under 12 Years* | | | | | |
|---|---|---|---|---|---|---|---|---|---|---|---|---|---|---|---|---|---|---|
| | *Kanam* | *Verumpattam* | | | *Others* | *Total* | *Kanam* | *Verumpattam* | | | *Others* | *Total* | *Kanam* | *Verumpattam* | | | *Others* | *Total* |
| | | *a* | *b* | *c* | | | | *a* | *b* | *c* | | | | *a* | *b* | *c* | | |
| *South Malabar* | | | | | | | | | | | | | | | | | | |
| Palghat | — | — | — | 43 | — | 43 | — | — | — | 3 | — | 3 | — | — | — | 3 | — | 3 |
| | | | | 100.0 | | 100.0 | | | | 100.0 | | 100.0 | | | | 100.0 | | 100.0 |
| Valluvanad | 160 | 8 | — | 36 | 19 | 223 | 24 | — | — | 13 | 4 | 41 | 29 | — | — | 26 | 8 | 63 |
| | 71.7 | 3.6 | | 16.1 | 8.5 | 100.0 | 58.5 | | | 31.7 | 9.8 | 100.0 | 46.0 | | | 41.3 | 12.7 | 100.0 |
| Ponnani | 136 | 1 | 3 | 41 | 41 | 222 | 132 | — | — | 40 | 13 | 185 | 130 | 2 | 47 | 85 | 26 | 290 |
| | 61.2 | 0.5 | 1.3 | 18.5 | 18.5 | 100.0 | 71.4 | | | 21.6 | 7.0 | 100.0 | 44.8 | 0.7 | 16.2 | 29.3 | 9.0 | 100.0 |
| Ernad | 238 | — | 2 | 19 | 5 | 264 | 129 | — | 2 | 17 | 4 | 152 | 76 | — | 2 | 56 | 11 | 145 |
| | 90.1 | | 0.8 | 7.2 | 1.9 | 100.0 | 84.9 | | 1.3 | 11.2 | 2.6 | 100.0 | 52.4 | | 1.4 | 38.6 | 7.6 | 100.0 |
| Calicut | 93 | 16 | — | 7 | 2 | 118 | 36 | 8 | — | 10 | 5 | 59 | 34 | 21 | — | 20 | 14 | 89 |
| | 78.8 | 13.6 | | 5.9 | 1.7 | 100.0 | 61.0 | 13.6 | | 16.9 | 8.5 | 100.0 | 38.2 | 23.6 | | 22.5 | 15.7 | 100.0 |

**Table 3.7** *Contd.*

| *Taluk* | *Above 30 Years but not Immemorial* | | | | | | *12 to 30 Years* | | | | | | *Under 12 Years* | | | | | |
|---|---|---|---|---|---|---|---|---|---|---|---|---|---|---|---|---|---|---|
| | *Kanam* | *Verumpattam* | | | *Others* | *Total* | *Kanam* | *Verumpattam* | | | *Others* | *Total* | *Kanam* | *Verumpattam* | | | *Others* | *Total* |
| | | *a* | *b* | *c* | | | | *a* | *b* | *c* | | | | *a* | *b* | *c* | | |
| *North Malabar* | | | | | | | | | | | | | | | | | | |
| Wynad | — | 24 | — | 31 | 7 | 62 | — | 49 | 2 | 57 | 1 | 109 | 4 | — | 103 | 172 | 3 | 282 |
| | | 38.7 | | 50.0 | 11.3 | 100.0 | | 45.0 | 1.8 | 52.3 | 0.9 | 100.0 | 1.4 | | 36.5 | 61.0 | 1.1 | 100.0 |
| Kurumbranad | 346 | 162 | 27 | 22 | — | 557 | 50 | 61 | 10 | 23 | 1 | 145 | 72 | 100 | 11 | 23 | — | 206 |
| | 62.1 | 29.1 | 4.8 | 4.0 | | 100.0 | 34.5 | 42.1 | 6.9 | 15.9 | 0.7 | 100.0 | 35.0 | 48.5 | 5.3 | 11.2 | | 100.0 |
| Kottayam | 93 | 56 | 34 | 5 | — | 188 | 17 | 54 | 20 | 1 | — | 92 | 17 | 99 | 24 | 1 | 2 | 143 |
| | 49.5 | 29.8 | 18.1 | 2.7 | | 100.0 | 18.5 | 58.7 | 21.7 | 1.1 | | 100.0 | 11.9 | 69.2 | 16.8 | 0.7 | 1.4 | 100.0 |
| Chirakkal | 1 | 92 | 427 | 18 | 8 | 546 | — | 47 | 235 | 76 | 8 | 366 | — | 41 | 218 | 92 | 11 | 362 |
| | 0.2 | 16.8 | 78.2 | 3.3 | 1.5 | 100.0 | | 12.8 | 64.2 | 20.8 | 2.2 | 100.0 | | 11.3 | 60.2 | 25.4 | 3.0 | 100.0 |
| Total | 1,067 | 359 | 493 | 222 | 82 | 2,223 | 388 | 219 | 269 | 240 | 36 | 1,152 | 362 | 263 | 405 | 478 | 75 | 1,583 |
| | 48.0 | 16.1 | 22.2 | 10.0 | 3.7 | 100.0 | 33.7 | 19.0 | 23.4 | 20.8 | 3.1 | 100.0 | 22.9 | 16.6 | 25.6 | 30.2 | 4.7 | 100.0 |

**Note:** Dash indicates nil.
**Source:** Tabulated from Government of Madras, 1882, Vol. 2, Appendix iv.

civil courts of Malabar from 1862 to 1880, Logan observed that by 1880 about one in every twenty cultivators had an eviction decree passed annually against him; that such decrees had more than quadrupled in the preceding 20 years; that the suit was, in fact, resorted to only for raising the rent, and it was usually only when the tenant refused to consent to an increase of rent that the decree was executed; and that the court cases were more numerous against the Mappilas than the Hindus (Government of Madras, 1882, Vol. 1:xxxvii–viii).

While these observations clearly confirm the recurrent theme of Logan's report that due to indebtedness, impoverishment, rack-renting, extortionate renewal fees, inadequate compensation paid at the time of eviction, and so on, the cultivating class was rapidly degenerating into a state of 'insolvent cotterism' (Government of Madras, 1882, Vol. 1:xxxii–viii), it was only too obvious to him that it was the thus widespread agrarian discontent which caused the Mappila outbreaks:

> [The] landlords are raising their 'rents' above their ancient customary shares of produce, and (eviction) is the readiest means which the law puts at their disposal for effecting that end. The *Mappilas* saw and realised this fact many years ago, and organised the *Mappila* outrages to put a stop to it. They decreed death against a landlord who evicted a respectable tenant, and they called in the aid of their religion to give its sanction to this 'pernicious doctrine' . . . . But the outrages were temporarily suppressed, and landlords, gaining confidence, have, in the last twenty years, resorted with increasing freedom to this means of bettering themselves. The recommencement of the outrages in the last decade, notwithstanding the enormous penalties imposed by the law, is, in my opinion, entirely attributable to this same cause. (Government of Madras, 1882, Vol. 1:lxxxi)

From Logan's inquiries into the conditions of the classes interested in the soil, it became evident that the one class whose interests deserved the most careful consideration was that of the actual cultivators of the soil, and that all other classes deserved to be regarded as mere investors of their money and as such perfectly competent to take care of themselves (Government of Madras, 1882, Vol. 1:xxiii). To redress the agrarian discontent of the district he felt it absolutely necessary to curtail the *janmi*'s power of *ouster* of the actual cultivator, and confer on the tenants the full benefits of the ancient customary law entitling them to sell the improvements of their holdings. To attain these ends he recommended the legislation of a statutory tenure for regulating the relations of the actual cultivator with his *janmi* (for details, see Appendix B to this chapter).

### *Madhava Rao Commission*

Logan's report, progressive as it may sound now, ran counter to the interests of the *janmis* and were hence unpalatable to the government. Therefore, in view of the intricacy of the subject and the interests involved, in January 1884 the government referred it to another Special Commission, with Raja Sir T. Madhava Rao as President and 4 others (including Logan) as members. This Commission also saw the political need for interfering with the *janmi*'s rights

> in the grave discontent prevailing among the agricultural classes regarding the growing insecurity of their tenures; in the fact that, among one class of the community, agrarian discontent, fanned by fanaticism (was) ready to develop at any moment into agrarian outrage; and in the existence of a system of tyrannies practised by some *Janmies* and apprehended from others. (Government of Madras, 1887:141)

In March 1884 the Commission submitted the draft Malabar Stay of Execution Bill, intended essentially as a temporary measure, and designed to maintain the existing relations between *janmis* and tenants, pending the revision of the law regulating those relations by the legislature (Government of Madras, 1887:140). This was followed by the draft Malabar Tenancy Bill, submitted in July 1884, along with its main report. Stating that it represented 'the historical claims of an historical class, namely, the hereditary raiyat or kudian,' the Commission contented that the *janmi*'s claim to a right of ouster against the tenant was unsupported by ancient usage or theory, and its draft Tenancy Bill proposed to limit it by giving an occupancy right to every tenant holding direct from a *janmi*:

> No tenant holding, direct from the *janmi*, the same land on any tenure not inferior to verumpattam for 30 years' continuously, or for 15 years, provided that the land was reclaimed within the last 30 years, or so holding through himself or another land which has been held on customary tenure (i.e., *kanam* or other superior tenure) for 30 years, and no tenant who may hereafter acquire the rights of such a tenant, shall be evicted except for wilful denial of his *janmi's* title, grave insult to his *janmi* or persistent refusal to pay rent: such are occupancy tenants; and no non-occupancy tenant, holding direct from the *janmi*, shall be evicted, except for denial of his *janmi's* title, non-payment of rent, waste or illegal transfer of his holding, until the determination of the tenancy. (Government of Madras, 1887:142, 154)

When the draft Malabar Tenancy Bill was referred by the government to the High Court for opinion, that, as well as Logan's report came under severe criticism. In a lengthy Minute Sir Charles Turner, then Chief Justice of Madras, held that Logan's inquiry was one-sided in method; that only the argument of the discontented was at all represented by it; and that the Commission to which his report was referred was composed of individuals who admittedly had formed their opinions before they joined, and who were all enlisted on one side of the question. Alleging that the *janmi*'s case had never been stated at all, he upheld the British administrative and judicial interpretation of *janmi* as full owner of the soil; and apprehended that the bill would multiply the already too numerous land tenures of Malabar, increase the dearth of land in the market, benefit one class and injure all the rest, and work a revolution in the tenure of landed property (Government of Madras, 1887:148–52).

While the reports and recommendations of the Logan and the Madhava Rao Commissions were criticised by several other persons as well, in Turner's strong dissent itself the government found sufficient pretext to postpone legislation: in September 1885, it appointed yet another Commission with C.G. Master, a member of the Executive Council of the Governor, as President, and 13 others including Madhava Rao, as members (Government of Madras, 1887:3).

### *Master Commission*

Unlike its predecessors, this Commission was of the view that 'the legislative recognition of an occupancy right in the tenants of Malabar was not justified either by historical considerations or in view of any political necessity'. At the same time, it admitted the need for protecting the tenants from capricious evictions. For doing this, in February 1887, along with its report, it submitted a draft Bill stipulating that no tenant shall be ordinarily evicted except at the end of the agricultural year and after six months' notice. It also admitted that 'a resort to legislation was necessary in order to secure to tenants, on eviction, the full value of their improvements,' and in February 1886 had submitted a draft Bill to that effect. The Commission had thought that these two Bills together, while involving a minimum of interference with the substantial rights of the *janmis*, would in some measure achieve the desired object (Government of Madras, 1887:4, 19–21). Of these two Bills, however, the government accepted only the compensation Bill, and enacted it as the Malabar Compensation for Tenants Improvements Act 1 of 1887. No doubt, it fell far short of Logan's recommendations. Nevertheless, it is significant as the first land reforms measure which the government was forced to legislate as a result of the Mappila outbreaks of nearly half a century. It also marked the beginning of a century-long history of land reform legislation in Malabar, mainly as necessitated by the protracted struggles and persistent demands of the various sections of the exploited peasantry.

The government hoped that this Act itself would go a long way in checking arbitrary evictions. But, many loopholes in it, especially the lack of clarity in defining compensation, the inadequacy of compensation awarded, and the misapprehension about the original intention of the Act caused by the many mutually inconsistent court rulings belied this hope. This was evident soon enough: reporting that the Act had neither given additional security to the tenants nor checked evictions and *melcharths*,[6] in 1893, Bradley, then Malabar Collector recommended legislation prohibiting *melcharths* and amending certain provisions of the Act (Government of Madras, 1928, Vol. 1:15). Though in 1900 the government repealed and re-enacted the 1887 Act as the Malabar Compensation for Tenants Improvements Act 1 of 1900, it also soon proved to be inadequate. For, as Ladejinsky wrote, 'what the "inferior interests" in land wanted was not so much compensation on quitting the cultivated land, which incidentally, was hardly ever granted in full, but a right to continue to cultivate the same land on payment of a reasonable rental' (Ladejinsky, 1977:76).

In 1911 when the government called for opinion on the working of the Act, Innes, then Malabar Collector was highly critical of it. To summarise his views: paddy lands were still as rackrented as ever and the act could do nothing to improve the condition of the tenants of such lands; though the Act was beneficial in respect of garden lands there were considerable difficulties in valuing improvements and complaints against valuation; the number of eviction suits was large; finally, the Act was not altogether fair to the *janmis*, the poorer of whom were sometimes compelled to have recourse

[6] *Melcharth*, also called *melkanam*, was a *kanam* given by the *janmi* to a third party with the power to redeem an outstanding *kanam*. It was created by a document known as *melcharth* and was treated as a mortgage. It operated for 12 years from the date of execution of the *melcharth* and not for 12 years from the date of redemption of the earlier *kanam*. The *melcharth* holder was entitled to add the actual costs of the action for redemption to the amount of the mortgage unless an amount was specified as being the amount payable in respect of the same. Though it was, in essentials, what property ownership generally entails, viz. the right to pass to another what belongs to oneself, no incident of the land tenures in Malabar had been the subject of more opprobrium than the *melcharth*. This was because of the abuse made of this power of transfer due to avarice or poverty of the *janmi*, wreaking vengeance on the tenant by the *janmi*'s underlings, and such other reasons. See, Government of Madras, 1928, Vol. 1:9, 68–70.

to *melcharths*. Following the remarks of the Board of Revenue that Innes' opinion was one sided and that it was not prepared to recommend legislation on such a thorny question without further inquiry, the government resolved to take up the question of legislation after further inquiry. Accordingly, it asked Innes to report whether he considered legislation feasible or desirable, and if so, to indicate the classes of tenants who required protection and the lines on which legislation should proceed (Government of Madras, 1928, Vol. 1:15–16).

In his report submitted in 1915, Innes pointed out that the 1900 Act had failed to achieve the principal object for which it was passed, viz. prevention of capricious and arbitrary evictions, that tenants were still entirely at the mercy of the *janmis*, that they were not better off under the intermediaries or *kanakkar* either, and that the actual cultivator of the soil had to be protected both against the *janmi* and the *kanakkaran*. He found ample grounds for legislation in the prevalence of rackrenting, arbitrary and capricious evictions, inadequate compensation for improvements, insecurity of tenure, exorbitant renewal fees, and the social tyranny of the *janmis*. Based on an examination of the statistics of population and the land available for cultivation, he concluded that it was a matter of economic necessity to give fixity of tenure to the tenant as the extension of cultivation would be greatly accelerated if the tenants who reclaimed lands were given more protection against eviction. He accordingly recommended fresh legislation conferring fixity of tenure on all cultivating and non-cultivating tenants who had been in possession of their holdings for 15 years and 40 years respectively. Evans, Innes' successor, to whom the government referred his report, stated that there were neither political nor economic reasons for legislation and that the legislation recommended by Innes would 'impose artificial and alien restrictions on a system evolved out of an accepted custom and would involve a reversion from a long established freedom of contract "to the tyranny of the law courts" '. The Board of Revenue which entirely concurred with Evans' views also stated that the legislation of the kind recommended by Innes would be a grave political mistake as it would alienate the *janmis* from the government. In the light of these views, the government again dropped its proposal for legislation (Government of Madras, 1928, Vol. 1:16–17; 1940, Vol. 1:15).

Meanwhile, the Mappila outbreaks, which began in 1836 and recurred at irregular intervals, culminated in the Malabar Rebellion of 1921. This mass insurrection lasted for about eight months, resulted in the death of thousands of people, and proved to be more violent and devastating than all the earlier outbreaks combined. Its immediate provocation was probably the relentless attacks on the Mappilas to suppress the Khilafat campaign fearing that it might arouse their anti-British and anti-*janmi* feelings. This was the view of both Kesava Menon and Madhavan Nair who were active in the nationalist and Khilafat movements in Malabar and in restoring peace and harmony after the rebellion (see, Menon, 1969:6–117; Nair, 1987). Its formative conditions were, however, more complex. Of them the most important was undoubtedly the deep-rooted agrarian discontent of the Mappilas. This was evident from their involvement in the tenancy agitations of the Nayars (more of this shortly), the Khilafat agitation,[7] and the non-cooperation movement of the Indian National Congress, all launched immediately before the outbreak of the rebellion:

> Had it not been for the agrarian discontent whose causes and extent had been correctly stated by Mr Logan half a century ago but whose remedy was not applied by the Government, the Moplah would not have rallied round the Khilafat and Congress flags in such large numbers and with such firm determination. It should also be remembered that most of the Congress and Khilafat leaders in

[7] In the First World War the attempts of the Allied Powers to dismember Turkey had deeply hurt the pan-Islamic sentiments since the Sultan of Turkey was also the Khalifa or Caliph—the spiritual leader of Islam. The tide of Muslim resentment turned anti-British through fear that the war with Turkey might lead to the liquidation of the Ottoman Empire and the Khilafat. The Khilafat (Caliphate) movement was therefore started by the famous Ali brothers, Muhammad Ali and Shaukat Ali, as a pan-Indian political agitation, for the restoration of the Khalifa and his temporal powers. See, Dale, 1980:180–83.

> Malabar (e.g., late K. Madhavan Nair, U. Gopala Menon, K.P. Kesava Menon, etc.) were also leaders of the Tenancy Movement. The two most controversial resolutions considered and ultimately passed at the Malabar District Political Conference held at Manjeri were those supporting Non-cooperation and Tenancy Reform. And it was after this conference that Congress and Khilafat committees sprang up in South Malabar in such large numbers as were not rivalled even in 1938–40. Every speech of every Congress-cum-Khilafat leader was a two-edged weapon aimed against both the British Government as well as the *Jenmi*. It was this that gave a new hope and new slogan to the oppressed Moplahs who joined the movement in such large numbers. It was a fine example of an agrarian political mass movement. (Namboodiripad, 1943:9)

These agitations merged with one another as is evident from the call of Ali Musaliar, one of the main leaders of the Khilafat agitation, and the rebellion for 'Khilafat, tenancy and Swaraj' as the panacea for the material problems of the poor Mappila peasantry (Dhanagare, 1983:78). They offered a political outlet to the agrarian discontent of the Mappilas for the first time in nearly a century of their struggles. The fifth District Conference held at Manjeri in the Ernad Taluk in April 1920 in which the first resolution demanding tenancy reforms legislation was passed, and which preceded the rebellion, was attended by a large number of Mappila tenants. In fact, in the first six months of the political agitations during 1920–21, it was tenancy reforms which attracted the most attention. Prompted by this, in January 1921, Thomas, then Malabar Collector, wrote that 'the non-cooperation movement hasn't caught on in Ernad where the Mappila is much more interested in the tenants' movement, and agitators can't get a hearing unless they make tenancy questions the big cry' (quoted by Dale, 1980:185).

During 1920–21 there were also growing attempts to organise the Mappila peasantry. The *Zamorin Estate Kudiyan Grievances Redressal Sangham* formed in 1920 near Kottakkal by the Mappilas themselves was perhaps the best example of these attempts. The tenants of this estate generally had complaints regarding exaction of excessive renewal fees and rent, when, following the eviction of a tenant from his holding, they resolved not to have that holding cultivated by any others. In one of the meetings of the tenants held at Kottakkal during this time not less than 5,000 Mappilas were reported to have been present, apart from K. Madhavan Nair and K.V. Gopala Menon, leaders of the *Malabar Kudiyan Sangham* (more of this *Sangham* shortly), and some other Hindus. Under the auspices of the *Zamorin Estate Kudiyan Sangham* mammoth tenant conferences were reported to have been held in a number of places. In January 1921 such a conference was held near Pukkottur, a hot-bed of the rebellion, attended by thousands of persons. The conference discussed both Khilafat and tenancy issues.[8]

After the crushing of the Malabar rebellion which marked the end of the first phase of peasant struggles in Malabar, the Mappilas withdrew to the background without having any major role to play in the subsequent phases of the peasant struggles. Some of the factors responsible for this seem to have been the brutal suppression of the Mappilas in 1921 by the British troops, failure of the Congress leadership to actively involve south Malabar in the subsequent phases of the nationalist movement for fear of igniting another rebellion, and the influence of Muslim League politics on the region.

## *Kanam* Tenants' Agitations and Land Reforms: 1920–30

While the peasant struggles in the first phase were carried on by the Mappilas (the bulk of whom were *verumpattam* tenants whose demands were mainly prevention of eviction and payment of compensation at the full market value), those in the second phase were spearheaded by the *kanam* tenants who belonged mostly to the Nayar caste, though like the

[8] See, for more details, Nair, 1987:91–96. It is not clear whether these activities were organised by the Mappilas themselves or were part of the Nayar agitations in the second phase of the peasant struggles. For other works on the Malabar rebellion, see Choudhary, 1977; Dale, 1980; Dhanagare, 1983.

former the latter were also stronger in south Malabar. While the Mappilas had waged their struggles all alone, lacking proper organisation, leadership, and ideology, the demands of the *kanam* tenants were backed by the nationalist movement. Moreover, most of the nationalist leaders themselves belonged to the Nayar caste, which, as already seen, occupied the middle stratum of the social structure and upper stratum of the peasantry.

The emergence of this phase of the struggles has to be understood in the context of the spread of English education in Malabar from the second half of the nineteenth century. Because of their privileges and pre-eminence in Malabar society, for a long time the Nambutiris considered English education an anathema, thanks to their false notions of aristocracy and social rank. This was, however, not the case with the Nayars, especially those from the *kanakkar* families. The new education opened to them the doors of British bureaucracy in Malabar, offered them its power and influence, and made them assertive and aspiring. Of the resulting conflict Namboodiripad wrote:

> The new class of educated young men and officers were politically and culturally far more advanced than their landlords who, however, were economically and socially dominant in the countryside. The very state which made them politically independent of the Jenmis made them much more dependent economically on those same Jenmis.
>
> A Thasildar or a Police Inspector or a Sub-Judge is part of a machine which deals with Jenmies as with any other citizen, but individuals who are appointed to these posts are socially and economically dependent on some of these Jenmies. The officer has innumerable opportunities of bossing over the Jenmies as over the rest of the people, but the Jenmi can evict his family from the house in which it lives. The educated and professional man with a wide outlook and a sturdy sense of self-respect has to humiliate himself before the narrow-minded and conceited ignoramus who is his landlord. (Namboodiripad, 1943:4)

This conflict had far reaching consequences. Among other things it provoked the Nayars to question the socio-economic dominance of the Nambutiris, and to condemn as a disgrace and fight against the *sambandham* of the Nambutiris with the Nayar women which was until then considered a pride and privilege by the Nayars. More importantly for the purpose of this study, it led to the emergence of the second phase of the peasant struggles, and provided it the necessary leadership mainly as attempts to protect the *kanam* tenants from exploitation, and oppression by the *janmis*. The exploitation and oppression by the *janmis* had taken the forms of arbitrary evictions, *melcharths*, and excessive levies at the time of *kanam* renewals over and above the customary dues. Instances of such levies were payment at the rate of one rupee per coconut tree and one rupee per *para* of paddy land, payment for *janmi*'s signature on the *kanam* deed (*oppavakasam*), as *janmi*'s nephew's right (*anandiravakasam*), and as mediator's right (*thandasthavakasam*) (Achutan, 1971:216).

The earliest of the attempts to protect the interests of the *kanam* tenants in the second phase of the peasant struggles was probably in 1912, when T.M. Nair, a member of the Madras Legislative Council (MLC) (and later one of the founders of the Justice Party), prepared a tenancy Bill to confer fixity of tenure on *kanam* and *verumpattam* tenants. As Nair resigned from the Council in 1913 his Bill had a premature death (Achutan, 1971:222). Subsequent attempts up to 1922 remained almost at the level of occasional meetings. The editor of *Mitavadi*, C. Krishnan, a Tiyya advocate and one of the few non-Nayar tenant leaders had recorded two such meetings: in August 1919 he met the new Malabar Collector and apprised him of the problems of the tenants. In October 1920 he participated in a meeting of the tenant leaders convened at Calicut at the residence of K.P. Raman Menon, a prominent lawyer and tenant leader, and contributed Rs 100 towards organisational expenses (Achutan, 1971:217). Between 1916 and 1920, the tenant leaders gradually captured the control of the Malabar Congress by ousting the *janmis* who dominated it till then. With this the demands for passing resolutions on the *janmi-kudiyan* (landlord-tenant) relations, voiced in each of the District Conferences starting from 1916 when the first District Conference in Malabar was convened, also gathered momentum (Nair, 1987:9–91). Thus, in the Manjeri conference mentioned earlier, K.P. Raman Menon pushed through a resolution demanding tenancy reforms legislation, despite stiff opposition from the *janmis* (Menon, 1972, Vol. 2:79, 81–82; Nair, 1987:91). In the following year, the first Kerala State Political Conference held at Ottapalam in the Valluvanad taluk demanded the harmonising of *janmi-kudiyan* relations (Pottekkat et al., 1978:415). In fact, practically in every regional political conference since 1920 when any significant political activity began to take place in Malabar, tenancy reforms figured as a major demand.

### Malabar Kudiyan Sangham

Meanwhile, when the Justice Party formed the first ministry in Madras in 1921 after the first election to the Reformed Council, some of the lawyers of the Pattambi Munsiff Court thought of finding a solution to the tenancy problem through legislative means. An immediate result of their efforts was the formation of the *Malabar Kudiyan Sangham* (MKS: Malabar Tenants Union) at Pattambi in 1922, with T. Rama Kurup, a prominent lawyer of the Pattambi bar, as president, and lawyers M.M. Kunhirama Menon, and P.A. Raman Menon as secretaries.[9] This was followed by the launching of the newspaper, *Kudiyan* (Tenant), with lawyer N. Kunhirama Pathiyar as editor. In December 1922 the MKS organised the first Malabar Kudiyan Conference at Pattambi. It was presided over by C.R. Reddy, a prominent member of the Justice Party and of the MLC, and attended by important tenant leaders. Among those present were another Justicite Mannath Krishnan Nair, and G. Sankaran Nair. The former had just entered the MLC after his Diwanship in Travancore. The latter had just entered the legal profession after giving

[9] After the Manjeri conference, K.P. Raman Menon and others formed the first *Kudiyan Sangham* at Calicut, with K.V. Gopala Menon, P.K. Kunhirama Menon, and C.K. Nair as secretaries. They decided to form *Kudiyan Sanghams* throughout Malabar, and in fact formed a few *sanghams* (Nair, 1987:91). The connection between these *sanghams* and the MKS is not clear. The former were probably part of the organisational efforts preceding the formation of the MKS.

up headmastership of a high school but with this Conference gave up his legal practice as well and became a full-time organising secretary of the MKS. The organisers of the Conference had given adequate publicity to their venture and to the oppression of the *kanam* tenants by the *janmis*. This Conference marked the beginning of the organised agitation of the *kanam* tenants.

The hard core of the agitation consisted of a few lawyers, intellectuals, and journalists from prominent *kanakkar* families in and around Calicut. Their methods of articulation were in keeping with their social and professional background. These involved the publication of articles and reports on tenancy issues in leading newspapers such as the Madras based *Hindu, Justice*, and *Mail* in English, and the Calicut based *Kudiyan, Mathrubhumi*, and *Mitavadi* in Malayalam; periodic meetings with the editors of the *Hindu* and the *Mail* for apprising them of the need for tenancy legislation; fighting elections on tenancy issues; seeking the patronage of important political leaders and inviting them to preside over the MKS conferences, and so on. Because of the fear of the tenants to participate in anti-*janmi* activities, initially meetings were held only clandestinely. But, due to persistent campaigning the MKS could organise about hundred of its units in different parts of Malabar in the course of the eight years of its agitation.[10]

### *MKS and MLC Election*

The election to the second Reformed Council in October 1923 witnessed the first active phase of the MKS. There was a special reason for this. Of the two persons elected to the Council from Malabar in 1921, Mannath Krishnan Nair, though himself a *janmi*, represented the *kanam* tenants. In his election manifesto he had promised to work for a tenancy Bill for Malabar and accordingly published one. His Bill proposed occupancy rights on all *kanam* tenants, both cultivating and non-cultivating, who had been in occupation of

[10] For details of the formation of the MKS, its early activities, and the Pattambi Conference organised by it, see Nair, 1971; also Achutan, 1971:217–20.

their holdings for not less than 25 years either directly or through predecessors. Though he presented the Bill in the Council in 1921 itself, it was not taken up for consideration till the Council was dissolved in July 1923 for re-election. Meanwhile, in December 1922 he had given notice of revising the Bill for conferring occupancy rights not only on all *kanam* tenants irrespective of the length of possession but also on all *verumpattam* tenants who were in possession of their holdings for six years or more. The widening of the scope of the Bill was obviously prompted by the prospects of the 1923 elections (see, Nair, 1971; Panikkar, 1978:886; also Government of Madras, 1928, Vol. 1:18).

Of the three candidates contesting this election claiming to represent the tenants, viz. Mannath Krishnan Nair, K.P. Raman Menon, and Ambattu Sivarama Menon, the first two were quite popular among the tenants. Apart from authoring the Malabar Tenancy Bill, Krishnan Nair had shown considerable interest in tenancy issues. In his election appeal he sought the support of the tenants for enabling him to get the Bill passed by the next Council. Raman Menon was mainly responsible for creating some awareness among the tenants and for forming the first *Kudiyan Sangham* in Malabar (see footnote 9). In his election appeal he stated that the main issue for the Malabar voters was the stipulation of *janmi-kudiyan* relations; he stressed the need for taking up the issue in the Council boldly and effectively, and assured that he would do so (*Mathrubhumi*, 11 September 1923). After several meetings and consultations and some confusion in the process, the MKS decided to back Krishnan Nair and Raman Menon. Though the MKS president Rama Kurup and secretary Sankaran Nair claimed that in doing this the MKS had ascertained the views of the tenant leaders from its different units, *Kudiyan* editor Kunhirama Pathiyar contested this claim and actively canvassed for the little known Ambattu Sivarama Menon (*Mathrubhumi*, 28 and 30 August 1923).

Of the three candidates, only Krishnan Nair was elected. While the defeat of Sivarama Menon was a foregone conclusion, the election of Raman Menon was rigged by the *janmis*. They had realised that this election was a trial of strength between them and the MKS, and that to retain their rights intact it was necessary for them to get the *janmi-sabha*[11] candidate Madhava Raja elected at any cost. As Sankaran Nair wrote, never before had they been so united, organised, and active (*Mathrubhumi*, 6 November 1923). Commenting on the election and the tenancy Bill, one *janmi* called for organising *Janmi Sanghams* to counter the *Kudiyan Sanghams*, for evicting the tenants and entrusting the lands to *janmis'* relatives, and for visiting each polling booth by the *janmis* and their men to keep an eye on the voting tenants (*Mathrubhumi*, 21 June and 9 October 1923). Apart from using the press and the platforms, the *janmis* resorted to intimidation tactics. Many *janmis* forced their tenants to vote for Madhava Raja and they did so. Many *janmis* waited at the polling booths watching their voting behaviour. Those who had doubts about the loyalty of their tenants either threatened them with eviction or prevented them from going to the polling booths (*Mathrubhumi*, 6 November 1923).

Despite the partial defeat of the MKS, the election provided it ample opportunities for building up a tempo on tenancy issues, both through the press and from the platforms. Though *Kudiyan* proved to be partial, *Mathrubhumi*—started just in time for the election propaganda with one *janmi* (Kurur Nilakantan Namboodiripad), five lawyers and one doctor as directors—devoted considerable space to lengthy editorials and debates on *janmi-kudiyan* relations and to election appeals by the MKS and its candidates. Every election meeting was also used for highlighting the tenancy issues. In fact, such meetings were relevant only insofar as they were focussed on tenancy issues.[12]

[11] Starting from the time of the Logan Commission, the *janmis* had begun to organise themselves to protect their interests as evidenced by some of the memorials submitted by them to the Madras Governor in 1886. See, GO 187, Legislative, 11 October 1886, and GO 191, Legislative, 19 October 1886. Their association, the *Janmi Sabha*, was probably formed then. Starting from 1904, the *Janmi Sabha* published a monthly, *Janmi*, edited by its secretary, Manavikraman Raja of Kottakkal, as the 'only organ of the Malabar landlords,' containing proceedings of its meetings, and articles on Malabar land tenures and land tenure legislations.

[12] For reports on the activities of the MKS during 1923 in the context of the election, see *Mathrubhumi*, March to November 1923; also, Nair, 1971.

### *Malabar Tenancy Bill and the MKS*

The period 1926–30 witnessed the most active and vocal phase of the MKS, centering around the Malabar Tenancy Bill and the government's opposition to it.

Krishnan Nair had begun his efforts to formally introduce his Bill in the MLC soon after the second Reformed Council was constituted. The Bill, eventually introduced in the Council in August 1924 and watered down considerably by a select committee, raked up the same old controversy on Malabar land rights. Most of the officials to whom it was referred questioned the need for such a measure and reiterated that the *janmi* was the absolute proprietor of the soil and that no case existed for giving fixity of tenure to the *kanakkar* (see, GO 366, Law (General), 5 February 1926). The main opposition was from the Law Member C.P. Ramaswamy Aiyar. In his 'open letter' of 1926 (of which more shortly), MKS secretary Sankaran Nair gave a lucid account of this opposition. To summarise part of it, albeit at the risk of simplification:

In the course of discussion when the Bill was introduced, Ramaswamy Aiyar made a speech on behalf of the government, comparable, for its equivocal character, to the Oracle of Delphi, though beyond pointing out certain 'difficulties' in the Bill and expressing general sympathy for homestead tenants and actual cultivators, he did not then make any definite statement on his government's views. The select committee headed by him dragged its feet for as many as fifteen months, though the number of days on which it actually transacted business was less than fifteen, and Ramaswamy Aiyar himself did not contribute in any way to its discussions. After the select committee signed its report he issued a dissenting note stating that the Bill as amended by it was so defective that it was incapable of being improved even by amendments. When his intention of thwarting the Bill became clear from his answers in the Council in February 1926, under pressure from the members the government changed its stance from total opposition to introduction of amendments in the Bill. But the task of framing these amendments was entrusted to F.B. Evans, its second secretary, whose anti-tenant views were so notorious that tenants throughout Malabar felt it necessary to protest immediately against his appointment.

According to Evans the tenancy agitation in Malabar was not due to any change in the economic situation that occurred since the government decided against tenancy legislation in 1918. He regarded the agitation as communal, Nayars vs Tiyyas in north Malabar, Nambutiris vs Samantans in south Malabar, and a combination of both in Calicut, due to the shape of party politics in Madras since the reforms (see, GO 530, Law (General), 14 December 1926). When he visited Calicut and held discussions with Sankaran Nair, it became only too obvious that his proposals cut at the very root of Krishnan Nair's Bill and were calculated to add insult to injury. Though prevention of arbitrary evictions was the chief cry of the tenants, there was not a single amendment to this effect (*Mathrubhumi*, 27 March 1926). When he reached Palghat he went straight to the residence of Raja Sir Vasudeva Raja of Kollengode, and held a conference of *janmis*, attended by many Nambutiris. The *janmis* had boycotted the select committee and insisted on maintaining the status quo. Upholding their views, he announced that he did not come across any evidence of deliberate evictions and harassment of the tenants by *janmis*, and that there was no need at all for any tenancy legislation (*Mathrubhumi*, 10 April 1926).

Infuriated by Evans' appointment and amendments, protest meetings of tenants were organised in different parts of Malabar. Of them a meeting held at Tikkodi in north Malabar in April 1926, presided over by the *Mitavadi* editor C. Krishnan and attended by tenants from different parts of north Malabar and tenant leaders from south Malabar, was probably the most significant. In organising it both the Nayars and the Tiyyas worked unitedly overlooking their caste differences. More importantly, it was organised with the help and cooperation of the younger members of *janmi* families in the area. Irrespective of its nature, their involvement was very significant, symptomatic of the larger changes in Malabar society under the influence of modern ideas and education. Apart from a lengthy editorial, *Mathrubhumi* reproduced

Krishnan's presidential address almost in one full page (see, *Mathrubhumi;* 20 March and 13 and 20 April 1926).

Meanwhile, the Bill was debated in the week-long meeting of the MLC in July 1926, when most of its sections were passed after rejecting Evans' proposals as they came up, after significant changes in the select committee's draft, and more importantly, after frustrating the attempts of Ramaswamy Aiyar to stall it. The remaining few sections were to be taken up in August 1926. Sensing trouble, Raja Sir Vasudeva Raja of Kollengode convened a meeting of *janmis* in July 1926, presided over by Kottakkal Valiya Raja, where more than a hundred *janmis* from different parts of Malabar participated. The meeting requested the Governor not to pass the Bill. Following this, a deputation of *janmis* led by the Kollengode Raja met the Governor at Ootacamund in August 1926, and secured an assurance from him to be very considerate to the *janmis* (*Mathrubhumi*, 5, 7, and 14 August 1926).

When the Bill was eventually passed by the Council on 2 September 1926, and when the tenants were still celebrating their victory by organising meetings in different parts of Malabar and felicitating Krishnan Nair and Sankaran Nair, came the news that the Governor had withheld his assent to it. According to him the Bill as passed by the Council contained inconsistencies, ambiguities, and other grave defects of form which would increase litigation and render it unworkable in practice if it became an Act; it was not profitable or just to go behind the judicial decisions of over three-quarters of a century which had recognised the full proprietary rights of the *janmi*; and any future attempt at legislation should take that right as a settled fact and proceed on the basis thereof (Government of Madras, 1928. Vol. 1:1).

The Governor's refusal to give assent to the Bill drew the ire of the tenant leaders and the tenants. On 1 December 1926, the MKS secretary Sankaran Nair issued a lengthy and learned 'open letter' (mentioned earlier) to the Governor, with copies distributed to the public, the Indian government, the British government, and members of Parliament, and reproduced in full in *Mathrubhumi* and *Mitavadi* as well. Stating unequivocally that ever since the Malabar Tenancy Bill was on the anvil the Madras government's policy towards it was neither just nor straightforward, he made a terse and telling survey of the events that transpired during the years 1924–26. He concluded his missive by reiterating the determination of the tenants to fight their battle to its logical end and to win their rights. In a letter Krishnan Nair appealed to the tenants for mobilising public opinion for the Bill through meetings, newspapers, letters to the government, meetings with members of the MLC, and men in public life, for starting a newspaper (*Kudiyan* was defunct by then), and for building up a 'Kudiyan Fund' for the purpose through contributions (see, *Mathrubhumi*, 7 December 1926; also Nair, 1971).

Meetings held in different parts of Malabar strongly protesting against the Governor's action, passed resolutions demanding the resignation of Ramaswamy Aiyar and recall of the Governor, and reaffirming faith in Krishnan Nair, Sankaran Nair and the MKS. Of them the most significant was probably the fifth annual conference of the MKS held at Mullassery in Ponnani taluk in February 1927, at the initiative of the *Mitavadi* editor C. Krishnan. It was presided over by L.K. Tulasi Ram, a prominent member of the MLC, and attended by about thousand tenants and tenant leaders from different parts of Malabar including Krishnan Nair and Sankaran Nair. There were as many as nine resolutions accompanied by lengthy speeches, thanking Krishnan Nair for his work for the tenants, accepting his appeals mentioned earlier, indicting the Governor, expressing lack of faith in the Madras government, and its Law Member Ramaswamy Aiyar, requesting Krishnan Nair to introduce another Bill without delay, and so on. Seconding one of the resolutions Sankaran Nair stated that in the history of Malabar tenancy two names would be remembered forever: that of Krishnan Nair as a selfless person who worked for the progress of the tenants, and of Ramaswamy Aiyar as an enemy of the Malabar tenants who worked for their ruin. Seconding another resolution, one of the Mappila tenant leaders, Kunhi Bapa Haji, recalled his warning at the Pattambi conference that if the government continued its anti-tenant policy there would be one more rebellion in Malabar. Apart from an editorial, *Mathrubhumi* published lengthy reports on the Conference on two consecutive days of about one page each out of its total space of eight pages (see *Mathrubhumi*, November 1926 to March 1927; also Achutan, 1971:224–25; Nair, 1971).

In the wake of mounting pressures, the Governor appointed a Committee in September 1927, with Diwan Bahadur T. Raghaviah Pantulu as President, and six others as members, to inquire into and report upon 'what disabilities, if any, (were) pressing hard on the tenants of Malabar in general; on the extent of unjustifiable evictions by the *Janmis* in particular, and how far the *Kanamdar* as such (was) in need of any protection' (Government of Madras, 1928, Vol. 1: 1–2). Since this Committee was filled with *janmis* and their supporters, the MKS protested against its appointment, gave a call for boycotting it, and campaigned against it throughout Malabar. In October 1927 it sent a deputation to Ootacamund and apprised the Governor of the absence of its nominee on the Committee (Achutan, 1971:24–26). In the face of total non-cooperation the Governor adopted a conciliatory attitude, added a tenants' representative as a member of the Committee, held discussions with a delegation of the MKS, and assured that he would hold a round-table conference of *janmis* and tenants to discuss the recommendations of the Committee. This was a clear victory for the MKS.

Meanwhile, there were some other happenings favourable to the tenants. In April 1928 Ramaswamy Aiyar's term as Law Member expired and in his place Krishnan Nair was appointed (see, Nair, 1971). Sankaran Nair went to Delhi, apprised the Congress Party high command of the tenancy situation and persuaded them to allow the Congress members to participate in the discussions on the Bill in the MLC by either opposing the government Bill or making necessary amendments in it. The consent thus secured was considered a great gain by the MKS.[13] More importantly, the attitude of the government towards the *kanakkar* had undergone a favourable change because of the fact that the agitation was championed by the educated middle class consisting of lawyers and government servants and the belated realisation of the political necessity of conceding to their demands. As a

[13] From the Diary of G. Sankaran Nair, MSS, Nehru Memorial Museum and Library, New Delhi, as quoted by Kunhikrishnan, 1985.

result of these happenings the subsequent events followed in quick succession.

The round-table conference assured by the Governor was held at Madras in 1929 from 7 to 10 and on 26 January, with seven *janmi* representatives and seven tenant representatives. It was inaugurated by the Governor and presided over by the Revenue Member and a tenant sympathiser Norman Marjorie Banks. Based on this conference, the government prepared a tenancy Bill and presented it in the MLC in August 1929, which after reconsideration by a select committee was passed by the Council in October 1929. While the Bill was awaiting Governor's assent, both the *janmis* and tenants made representations to him through deputations and other means (see, GO 4111, Law (General), 23 October 1929). After further modifications by the Council as suggested by the Governor, the Bill got his assent in March 1930 and was sent for the Viceroy's assent. Not to be outdone, the *janmis* engaged Ramaswamy Aiyar to represent their case. The MKS leaders lost no time in protesting against the unfairness of allowing a former Law Member with considerable influence on officials to lead a deputation on behalf of the *janmis*. Following this, the *janmis* were forced to engage another person to lead their deputation. Sankaran Nair stayed at Simla for seven months countering the machinations of the *janmis*. Though the Bill was returned twice to Madras due to pressures from the *janmis*, after hearing the deputations of both the *janmis* and the tenants, the Viceroy eventually gave his assent to it. The Malabar Tenancy Act XIV of 1930, which came into force on 1 December 1930, was the outcome (see, Menon, 1973: 152; Nair, 1971).

While the enforcement of this Act marked the close of the second phase of the peasant struggles in Malabar, the events in this phase raise a number of important issues. Malabar is one of the few places in India where land relations were intensely and effectively articulated by an educated middle class as early as in the first quarter of this century. Though the British administration stood stolidly by the *janmis* throughout nearly a century of agitations by the illiterate, impoverished, and inarticulate Mappilas, when the educated, affluent, and articulate Nayars appeared on the scene, in less than a decade it conceded to their demands as a matter of political expediency. For, the main beneficiaries of the 1930 Act were the *kanakkar* themselves:

> Almost all the demands which the Kanamdar had been making for half-a-century have been conceded. He can no more be evicted by a capricious Jenmi at his sweet will and pleasure; his interests in the land get as much value as the Jenmi's own; the professional and commercial people who accumulate decent amounts seek him out and try to purchase his interests thus converting him into the possessor of almost liquid cash. His rents can't be enhanced, nor can the Jenmi make arbitrary levies on him as he used to do. The renewal fee that he was to pay to the Jenmi every 12 years had lost all the terrors of a periodically erupting volcano ruining his entire family as it could now be correctly calculated according to well-known and fixed principles in contrast to the arbitrary and unknown whims of the Jenmi, as hitherto. (Namboodiripad, 1943:12)

After the introduction of this Act the attitude of the *kanakkar* towards the *verumpattakkar* was similar to that of the *janmis* to the tenants when the former were granted full ownership rights by the British authorities. In the subsequent years the *kanakkar* became what Ladejinsky termed as 'the symbol of typical landlordism' in Malabar (Ladejinsky, 1977:76). Thus, the passing of the Act while curbing the rights of the traditional *janmis* created a new class of *janmis*.

Though the Act provided for fixity of tenure and fixation of fair rent in the case of the *verumpattakkar* as well, because of the many loopholes in it (especially the *janmi*'s right to eviction for failure to pay one year's rent in cash as security and for personal cultivation) made these provisions mostly illusory to all but the few rich and powerful *verumpattakkar* (Namboodiripad, 1943:14). The agitation of this phase was not a mass movement in the full sense of the term, inasmuch as the overwhelming majority of the *verumpattakkar* were kept out of its purview and the demands were mainly confined to those that concerned the *kanakkar*. Nevertheless, it was this agitation which gave the Malabar peasants 'the first elements of class consciousness—the consciousness that they should unite as a class and fight their enemy, the *Janmi*' (Namboodiripad, 1967:112). This class consciousness was put into practice in the third phase when the peasant struggles were carried forward mainly by the vast masses of the *verumpattakkar*, who in the second phase remained the meek camp-followers of the *kanakkar*.

## *Verumpattam* Tenants' Agitations and Land Reforms, 1930–56

In contrast to the first two phases, the peasant struggles in the third phase were stronger in north Malabar. In this phase the struggles were no longer confined to an interest or caste group, but took the nature of a well-organised class struggle, with interests of the lower peasantry at the fore and the *verumpattam* tenants in the forefront. As the struggles got politicised they were transformed into a broad-based social movement directed against landlordism and the caste system on the one hand, and imperialism on the other. The role of *verumpattam* tenants belonging to the lower strata of Hindu society, in particular to the Tiyya caste, became especially spectacular in this fight against social evils. Perhaps the close nexus between the economic and social conditions and the difficulty of improving the former without alterations in the latter and vice-versa had become apparent.

For the emergence of the organised peasant struggles in this phase, agrarian discontent was only a necessary but not a sufficient condition. Without firm and effective leadership and a specific ideological direction, it would have been hardly possible to channelise this discontent and give it an organised expression. The large-scale evictions, rack-renting, and indebtedness caused by the steep fall in the prices of agricultural commodities, especially of commercial crops, during the depression years 1929–33, and the steady rise in the revenue assessment rates due to the resettlement undertaken in 1929 led to mounting discontent (Karat, 1976:31) and rendered the conditions for organising the peasantry ripe and favourable. While the nationalist leaders released from jails after the suspension of the Civil Disobedience Movement in 1933 gave a call for organising the peasants and workers, a new type of Congress cadre which the nationalist struggles of that period gave rise to and which in May 1934 formed into the Kerala Congress Socialist Party (KCSP)[14] resolved to champion their cause. Of the emergence of these cadres and their development into the leadership of the peasant struggles of the third phase, Namboodiripad wrote:

> The social composition of the Congress Volunteers in Malabar was predominantly rural middle class . . . . (Apart) from a few individuals from among the progressive Jenmi and bourgeois elements and also a number of nationally conscious (but not class conscious) workers, the great majority of Satyagrahis were connected with land as tenants. They were, however, junior members of their families, who had some elementary education, were either unemployed or had some job as local school teachers . . . .
>
> The course of the national struggle, outside the heated discussions over it inside the jail, the earnest study of political questions and the close contact with some revolutionaries, made these Satyagrahis into convinced revolutionaries by the time they came out of jail . . . . (This) proved adequate material for the building up of the Congress Socialist Party. Almost the entire ranks of the Satyagrahis joined the Party and became its active workers. And their social composition made it inevitable that they turn their main attention to the peasant movement . . . .
>
> Every Satyagrahi, on coming out of jail, set himself up in his own village. He started a small reading room where the young men of the locality would gather every day, read the daily paper and discuss politics. He would also tell them what he learned in jail, whom he met, what

[14] KCSP was the counterpart of CSP and a ginger-group of the Congress consisting of persons with a socialist stance who had become disillusioned with the mild-Gandhian means of fighting the British and realised that the nationalist movement could become strong and sustained only when the peasantry, who formed the bulk of the rural population, were drawn into its mainstream. Their attempt was therefore to link the peasant struggles with the nationalist movement.

they talked about, what they had decided to do. This reading group gradually grew in number and this reading room became ever more and more the centre of political activity . . . . Everywhere the youth in North Malabar was showing its mettle as an organiser of itself which was later to come out as the new organiser and leader of the new peasant movement. (Namboodiripad, 1943:15–16)

### *Emergence of Karshaka Sanghams*

Meanwhile, a public meeting held at Calicut in November 1933, presided over by Manjeri Ramaiyar, and where about fifty persons participated, including U. Gopala Menon, M. Govinda Menon, T. Hassan Koya Mulla, K.A. Keraleeyan, P. Krishna Pilla, K. Madhava Menon, and E.M.S. Namboodiripad, resolved to form the *Kerala Karshaka Sangham* (Kerala Peasants' Union) in the whole of Malabar, with taluk and village *sanghams* as its units for safeguarding the interests of the peasantry and to immediately apprise the government of their hardships due to the increase in revenue rates. It also formed a working committee authorised to enrol more members (*Mathrubhumi*, 7 November 1933). This was followed by an active tax-remission campaign at tax-payers' and peasants' conferences as the beginning of organised peasant struggles which soon struck deep roots at the village level, and which tried to put an end to the exploitation and oppression of the peasantry (see, Karat, 1976:31).

The exploitation of the peasantry had taken several distinct forms besides rack-renting, extortionate renewal fees and the constant threat of eviction. The *janmis* had devised various other arbitrary and illegal exactions like *vasi*,[15] *nuri*,[16] *mukkal*,[17]

[15] *Vasi* was a margin to make up the deficiency in the quantum of grains through drying. This was three *paras* (*para* is a measure of roughly 10 kilograms) for every 10 *paras* measured out as the *janmi*'s share.

[16] *Nuri* literally means what three fingers can hold. This was a large pick of paddy kept aside for every 10 *paras* of paddy measured out as the *janmi*'s share, as a means of numbering.

[17] *Mukkal* means three-fourth. This was a levelled *idangazhi* (a measure of roughly 1 Kilogram) of paddy given to the *janmi* after measuring out all his share, for filling up the likely shortage.

and *moonnidangazhi*,[18] and feudal levies like *vechukanal/seelakkasu*,[19] *kazhcha*,[20] and *kankani*,[21] which had trussed up the peasantry into an economic straitjacket. Socially, members of the polluting castes were the worst affected as they were also subjected to severe social disabilities. Under these conditions, how long a tenant held his land depended on how promptly and how best he appeased his *janmi*'s greed and cupidity. The *janmi* had several traditional weapons in his armoury to humiliate and persecute the tenants and these included the denial of essential services like those of the barber, washerman, priest, washer-woman's purificatory cloth (*mattu*) to women, and the use of the village pathways and wells. He could also order his men to flog recalcitrant tenants, rob or ruin their orchards, steal their livestock, set fire to their hayricks, or molest their women. It was to fight against such social oppression and exploitation that *Karshaka Sanghams* (KSs) emerged at the village level from 1935, which spread to the whole of Malabar in the next three years. With the formation of the *All Malabar Karshaka Sangham* (AMKS) in 1937 the KS became a three-tier organisation with the village KS as the smallest unit, the taluk KS above it, and the AMKS as the apex coordinating body.

The progress of the peasant struggles during the years 1935–40 was closely bound up with the emergence of the KS as a strong peasant organisation. In organising its activities the KS fully utilised the prevailing discontent among the peasants and their opposition to the *janmis* and the government. Massive peasant conferences at the village, taluk and

[18] *Moonnidangazhi* means three *idangazhi*. Three *idangazhi* of paddy were offered as wages to the *janmi*'s measurer.

[19] *Vechukanal/seelakkasu* was an offering of money made to the *janmi* before seeing him every time, as a peasant approaching his *janmi* without this was contrary to customary behaviour.

[20] *Kazhcha* was an offering made to the *janmi* on special occasions like birth, death, festivals, and ceremonies in the *janmi*'s house. This was usually in the form of headloads of fruits and vegetables.

[21] *Kankani* literally means the offering of first fruits to the *janmi* when the harvest was about to be reaped. The custom prevalent was to offer one bundle of betel leaves, 10 arecanuts, and one tin of ghee to Hindu *janmis*.

regional levels, and processions of peasants and youth[22] singing revolutionary songs and shouting rousing slogans were more or less regular features of its programmes. The demands of the peasantry remained more or less the same throughout: these included ceiling on rent, abolition of feudal levies and illegal exactions, use of standardised measures instead of fake ones, and amendment of the Malabar Tenancy Act of 1930 in a comprehensive way.

Massive intercaste dinners, for which the peasants carried headloads of vegetables and rice, were a unique and recurrent feature of the conferences organised by the KS, which went a long way in weakening intercaste barriers and promoting class solidarity. Staging of plays with revolutionary themes often highlighting the evils of landlordism and exploitation of the peasantry[23] was another. It is therefore hardly surprising that in north Malabar, where peasant struggles were the most powerful, caste distinctions have all but disappeared except in the case of the Nambutiris and the Harijans. Incidently the role of these polar groups in the peasant struggles was also minimal for the obvious reason that the former were *janmis* while the latter formed the bulk of the landless labourers.

The formation of the *All India Kisan Sabha* at Lucknow in April 1936, which had as its aim the complete emancipation of the peasantry from economic exploitation, gave a fillip to the peasant struggles in the whole of India (Pothuval, 1976:44). Following this, hunger-*jathas* were organised in different parts of the country. Of the hunger-*jathas* organised in Malabar, the most popular was the one taken out by A.K. Gopalan from Cannanore to Madras in July 1936 (see Gopalan, 1976:8–84). This was followed by taluk-level *jathas* organised in Chirakkal, Kottayam and Kurumbranad during September-October 1936 (see, *Mathrubhumi*, 24 September and 15 October 1936). The hunger-songs sung by peasant

[22] Youth *sanghams* like A.V. Kunhambu's *Abhinava Bharat Yuvak Sangham* at Karivellur, and K.P.R. Gopalan's *Youth League* at Kalliasseri were active in organising the youth and peasants even before the KS came into being.

[23] *Pattabakki* (rent-arrears) and *raktapanam* (the drink of blood) by K. Damodaran, and *koottukrishi* (collective farming) by Edasseri were the most important of these dramas.

activists K.P.R. Gopalan and K.A. Keraleeyan had profound mass appeal.

The first Chirakkal Taluk Peasants' Conference organised at Parassinikkadavu in November 1936 at which the Chirakkal Taluk KS—the first taluk KS in the whole of Malabar—was formed, was a major event in the peasant struggles of this period. Presided over by barrister A.K. Pilla, and attended by peasants from different parts of the taluk such as Karivellur, Bakkalam, Cherukunnu, and Kalliasseri, who marched to the venue in well-organised *jathas*, this Conference called for starting a KS in every village and enlisting a member for it from every peasant family. It resolved that *janmis* were entitled only to that part of the produce of the soil which remained after meeting the requirements of the peasants for their living. It urged for the introduction of effective legislation to this end and till such time for adopting certain ameliorative measures. These were fixation of rent at one-fourth of the net produce, that is, after deduction of the cost of cultivation, prohibition of the collection of all levies other than rent, security of tenure and cancellation of rent arrears. It also expressed its concern for fixing wages of agricultural labourers and arresting the fall in the agricultural prices (*Mathrubhumi*, 4 November 1936).

Following this Conference numerous peasant conferences were organised at the village, taluk, and regional levels throughout Malabar, including Kasargod taluk which was then part of South Kanara (see Karat, 1976:36–37; Pothuval, 1978:21–22). During the short period of five months between January and May 1939, as many as thirty-three village conferences were reported to have been organised (Balaram, 1973:83). The growth of the KS during this period was phenomenal. By the end of 1938, the AMKS had a paid up membership of about 30,000—5,000 in Kasargod taluk, 10,000 in Chirakkal taluk, and the rest in the rest of Malabar with Kottayam and Kurumbranad taluks probably accounting for the bulk of it (Jeffrey, 1978:91). By the end of the same year, there were more than 150 village KSs and a taluk *sangham* in each taluk (Balaram, 1973:44).

### *Congress Betrayal and Peasant Response*

When the Congress Party backed by the peasants' and workers' organisations won the provincial elections and formed a ministry in Madras in July 1937 under C. Rajagopalachari, its pretension of championing the cause of the peasantry, as announced in the Faizpur session of the AICC in 1936 and in its election manifesto, was fully exposed. With great expectations from the Congress and pinning their hopes on it, the peasant leaders engaged in propaganda work for the Congress and exhorted peasants and workers to join it for strengthening the mainstream of the nationalist movement. But the ministry failed to provide any relief to the peasants other than the mere passing of the Agricultural Debt Relief Act. This betrayal on the part of the Congress leadership disappointed the leftists within its ranks and alienated the entire body of peasantry which soon came to realise that any improvement in their conditions was possible only through their own organised strength. Accordingly, they resolved to launch a two-pronged struggle against the *janmis* on the one hand, and the *janmi*-backed Congress government on the other. Massive and militant *jathas* were taken out throughout Malabar to the houses of almost all prominent *janmis* (see, Karat, 1977:5–8; Kurup, 1978:12–13; Pothuval, 1976:23–28, 40).

Apart from a strong and sustained campaign against payment of rent to the *janmis* and revenue to the government, social boycott which had until then been a powerful weapon in the hands of the *janmis* was now wielded by the peasants effectively and applied against the recalcitrant *janmis* and, for that matter, any villager who did not support the KS programmes. Alarmed by these agitational tactics, the South Kanara Collector in a message directed the people to report to the *potail* (village headman) any action of social boycott or violence by the KSs. The *potail* was instructed to send daily reports on such cases to the nearest police station. Towards the end of 1938 he also visited the Kasargod taluk (and the Malabar Collector the north Malabar taluks), for first-hand information on the situation (*Mathrubhumi*, 22 and 27 December 1938). A *janmi*-conference held at Kottayam palace near Tellicherry in January 1939 requested the Congress authorities to take appropriate action against those Congressmen who exhorted the peasants to the path of agitation (*Mathrubhumi*, 26 January 1939). This phase of the peasant struggle, characterised as it was by the defiance of customary authority and by direct action, was a source of considerable concern to the British administration as is evident from a despatch:

> The agrarian situation in the Chirakkal and a portion of the Kottayam Taluks of Malabar and the Kasargod Taluk of South Kanara continues to cause some anxiety. The agitation is based on village Karshaka Sangams and Sub-Sangams, composed largely of tenants and labourers in which some teachers of the District Board and aided schools are reported to have taken part. Longstanding grievances of the tenants against the Janmies are worked up, and lead to demands for the abolition of a number of customary payments, the drastic reduction of rents and the use of standardised measures: Recalcitrant jenmies and others who have incurred the displeasure of the Sangams, are subjected to social ostracism and boycotting by the barbers, washermen, etc. It is reported that as a consequence the respect for law and order has been affected.[24]

[24] This dispatch is one of the fortnightly confidential reports sent by the Public Department in Madras to the Home Department in Delhi for the first half of December 1938. To cite a few more of the abstracts from similar reports:

'In Malabar there has been for sometime an agitation urging tenants not to pay their dues to the jenmies (landlords). The District Magistrate reports that generally speaking the tenants are not paying' (from the report for the first half of August 1938).

'According to reports received from the District Magistrate of South Kanara, the Kasargod Taluk Peasants' Association and Malabar Socialists have been asking the people not to pay land revenue and at the time of attachment and distraint to cause as much obstruction as possible to the authorities' (from the report for the second half of October 1938).

'The District Magistrate of Malabar reports that a no-rent campaign (among lease-hold tenants in respect of dues payable to the (ryotwari Pattadars) which has been carried on for some time in that District, is achieving considerable success, and that in the absence of any organised

The organised struggles of the peasantry questioning the age-old feudal oppression and exploitation forced the *janmis* to give up feudal levies, illegal exactions, and also to switch to the use of standardised measures. This was a major victory for the peasants and it brought them a new sense of strength.

### *Militant Agitations for Malabar Tenancy Act Amendment*

At the same time, the agitation for the amendment of the Malabar Tenancy Act of 1930 also tended to be more and more militant. On 1 September 1938 which was observed as All-India Peasants' Day, the KSs throughout Malabar passed resolutions in support of the amendment. They observed 6 November 1938 as Malabar Tenancy Bill Amendment Day when a uniform resolution reiterating their earlier demands was passed. On this occasion, a draft Bill incorporating all the amendments demanded until then was printed and circulated (Karat, 1977:7–8). This was in reaction to the notice given by the government in October 1938 of the introduction in the Madras Legislative Assembly of a Bill to amend the Malabar Tenancy Act of 1930, with its scope restricted to the removal of certain difficulties in the working of the Act which evidently did not satisfy the peasantry. Based on a report submitted by a sub-committee appointed by it, the KPCC also appealed to the government for enacting a more comprehensive Bill (Pottekkat et al., 1978:547–48). The Malabar Tenancy Amendment Day was followed by an All Malabar Peasant Rally in December 1938 which saw the

---

opposition is in some parts undermining the authority of the Government. He fears that, if the jenmies are unable to collect their rents, it will have a serious effect on the land revenue collections which start next month. There is a rather similar agitation reported from the southern part of South Kanara, where activities of the Peasants' Union have led to some disturbances. The Government is considering the action which should be taken to counter this development' (from the report for the first half of November 1938).

'There are no signs of any improvement in the agrarian situation in North Malabar. Reports have been received of several intemperate speeches in which complete disregard to the jenmies is urged and reference is made to the small number of the police in the Police Stations of the interior' (from report for the second half of November 1938).

march of two *jathas* each of five hundred red-clad peasants to Calicut, representing the southern and northern regions of Malabar, and a third *jatha* of a hundred and fifty persons to Mangalore, representing the Kasargod taluk. At the end of these *jathas* memorials containing the grievances of the peasantry were presented to the respective district collectors. The merger of Kasargod with Malabar and the extension of the Malabar Tenancy Act to Kasargod were also demanded at the Kasargod *jatha* (see Gopalan, 1976:96; Karat, 1977:8; Kurup, 1978:14, 88–93).[25]

In view of the mounting tenant agitations and the persistent demands for a more thorough legislation, the government did not introduce the Bill for amending the Tenancy Act of 1930. Instead, it decided to consider the necessity for legislation of a more comprehensive nature, and deputed its revenue minister T. Prakasam to visit Malabar to study the situation. The KS presented to him copy of a memorandum sent by it to the Malabar Collector and of a draft tenancy Bill prepared by it (*Mathrubhumi*, 29 December 1938). Based on his report, in July 1939 the government appointed a 'non-official committee' with K. Kuttikrishna Menon as chairman and eight others as members to inquire into the working of the Malabar Tenancy Act.

But, as observed by *Mathrubhumi* in an editorial, there

[25] Of these *jathas*, abstracts from the fortnightly reports: 'Jathas are being taken to Calicut for a 'conference,' and a deputation is being arranged for the purpose of presenting to the Collector a list of grievances. It is expected that he will receive it on the 18th or 19th. His offer to receive such a deputation while he was camping in the affected area was not accepted. The organisers insist on his receiving it in Calicut, where it can be expected to attract the most publicity, and to impress on the public the standing and influence of the leaders' (from the report for the first half of December 1938). 'Interest aroused by the agrarian agitation in Malabar centered on the Peasants Conference held in Calicut on the 18th. The opening speech by a Congress MLC who put the moderate view and advised the abandonment of the weapon of social boycott, is reported to have damped the ardour of the more militant delegates from the north of the District. Resolutions dealt mostly with agrarian subjects, though one advocated the establishment of a Socialist State. The Collector was unable to meet a deputation at the time of the Conference, but has arranged to receive one in January. The Honourable Minister in charge of Revenue Department has been touring in the District' (from the report for the second half of December 1938).

was some kind of *janmi-kanakkaran* alliance in the Committee. For, while their interests were fully represented, those of the *verumpattakkar* were not (*Mathrubhumi*, 19 December 1940). The Committee had on it two avowed *janmis*, both *janmi* representatives in the Madras Legislature, and a dozen others who represented the *kanakkar*, but not a single MKS representative, though three of the MLAs appointed to it (viz. E. Kannan, a Harijan, KPCC President Muhammad Abdur Rahman Sahib and Secretary E.M.S. Namboodiripad) strongly supported the peasantry and the MKS (Namboodiripad, 1943:28). Though the MKS was dissatisfied with this composition, as Namboodiripad wrote, it decided to use the Committee as a lever for building up a mighty mass movement:

> At every centre visited by the Committee, there were huge rallies and demonstrations welcoming the committee and urging on it to concede their (peasants') demands. While among the witnesses who tendered evidence before the committee formally, there were only one or two representatives of Karshaka Sanghams, the actual number of peasants who supported these witnesses were not less than 25,000. (Namboodiripad, 1943:28)

Though the Committee submitted its majority report in December 1940, its recommendations turned down the main demands of the peasantry, such as abolition of renewal to the *kanakkar*, abolition of one year's rent as deposit by the *verumpattakkar*, and reduction in rent rates (Namboodiripad, 1943:29). Not surprisingly, the three members supporting the peasantry struck discordant notes. Of them, the dissenting note by Namboodiripad is still a valuable document on the evils and irrelevance of landlordism as a social institution and strongly pleaded for its abolition as a precondition for any development. He dwelt at length on certain basic questions of land tenure, such as 'whether landlordism as an institution serves any useful social function or whether it is parasitic in nature, whether its continuance is a necessity for society at large, or whether it should be ended with or without compensation' (Government of Madras, 1940, Vol. 1:71–84).

With the outbreak of the Second World War, and the resignation of the Congress ministry on the ground that 'India had been made to participate in the war without her consent',[26] and also due to pressures from the *janmis*, the government shelved the report for more than a decade. Meanwhile, in December 1945, it amended the Act of 1930 limiting the conditions of eviction to purposes of self-cultivation and the construction of buildings, one at a time (see *Mathrubhumi*, 7 and 15 December 1945). But, on the pretext of self-cultivation, the *janmis* evicted tenants and expropriated either the competitive rent by leasing out the land to new tenants or the high price of paddy in the black-market through direct cultivation.

The crisis that was brewing within the Congress Party since 1937 came into the open with the outbreak of the Second World War. The radicals in the Party were not satisfied with the mild protest of the CSP which called upon its members to send to the district collectors individual postcards opposing the war. The result was a secret meeting of about ninety KCSP workers in December 1939 at which the Communist Party in Kerala was formed, though it did not formally break away from the KPCC for some time to come (Andalat, 1979:361–62).

### *Repression and Reprisal*

The refusal of Britain to give an assurance to grant India independence after the war disappointed the freedom fighters throughout the country. In Malabar the communists announced their intention to observe 15 August 1940 as 'Civil Liberties'/'Anti-Repression Day,' though they could not do so because of a ban on meetings and processions that day imposed by the District Magistrate.[27] Following the ban, they gave a call for observing 15 September 1940 as 'Protest Day'. Defying the ban order this time, they organised public meetings and processions in several places. The attempts of the police to disturb and disrupt these meetings and

[26] See fortnightly report for the second half of October 1939.

[27] See fortnightly reports for the second half of August 1940, and first half of September 1940.

processions resulted in police-public confrontations and clashes in several parts of Malabar—such as Morazha, Mattannur, Payyanur, Cannanore, Tellicherry, Badagara, Trichur, Pattambi and Malappuram. The immediate results were the death of a police sub-inspector and head constable at Morazha village in Chirakkal taluk, and two KS volunteers, Abu and Chathukutty, at Tellicherry town in Kottayam taluk, and the injury and arrests of many in different places (*Mathrubhumi*, 16 September 1940).[28]

Following the violent incidents of the Protest Day, the police force in Malabar turned trigger-happy and unleashed assaults and atrocities on whoever they thought were against the British, spreading a wave of terror throughout the region. While the leaders of the KS had either gone underground or been put in jail, the communists were determined to retaliate and pull down the already shaky British administration and along with it throw off its pernicious ally, landlordism. The historic 'Kayyur Riot' was but one manifestation of this resolve. It originated in a clash between a police constable and a group of villagers on 28 March 1941, in which the former was allegedly stoned to death in a river. This led to a long and terrifying police manhunt[29] in a number of villages and the arrest and trial of sixty persons, of whom eighteen were sentenced to imprisonment of varying periods and four (viz. Abubacker, Madathil Appu, Koyithattil Chirukantan and Potavara Kunhambu Nair) were hanged to death at the Cannanore central jail on 29 March 1943 (see, *Mathrubhumi*, 30 March 1943).[30]

[28] On this the fortnightly report for the first half of September 1940: 'serious riots occurred at five places resulting in the death of a police sub-Inspector and one constable and serious injuries to others at Pappinisseri and in the firing by the police at Tellicherry where two persons were killed. There were lathi charges in the other places.'

[29] According to the fortnightly report for the first half of April 1941, 'after the crime (had) been committed all those who had taken part fled to the surrounding jungle and it (had) been necessary to call out the Reserve Police and three platoons of the Malabar Special Police to comb out the jungle in search of the culprits'.

[30] The Kayyur riot is an important event in the history of peasant struggles and nationalist movement in Kerala. It has been the subject of an excellent novel in Kannada by one of its eminent writers (Niranjana's *Chirasmarane*); an exhaustive account in Malayalam by a frontline peasant and communist

While the events of the Protest Day resulted in the dissolution of the KPCC by the AICC and the appointment of an ad hoc committee, the Kayyur riot brought an indefinite ban on the AMKS and all its units. However, the formation of the *All Kerala Kisan Sangham* in 1942 with representatives from Travancore and Cochin areas as well clearly proved the inability of the government to crush the peasant struggles.

After the war, when the Congress Party came to power in Madras in April 1946 with T. Prakasam as prime minister, instead of bringing succour to the suffering peasantry it tried to crush the peasant struggles in Malabar, lest the leftist insurgency should engulf it. The *Kisan Sangham* was not disheartened by this. Its immediate concern was to combat the famine that was still raging. It designed two main strategies to do this: to curb hoarding and blackmarketing by preventing the movement of foodgrains from one area to another and by getting the surplus levied at the fair price shops; and to cultivate all the available dry and waste lands belonging to both the government and the *janmis*, with a view to raising the agricultural output. It also gave a call to the peasantry for forced occupation of these lands if the government failed to make the same available to them before 15 December 1946. The attempts by the police and the *janmis* to thwart these schemes triggered a series of violent incidents in different parts of Malabar resulting in the death of at least eight persons in two places. The wave of repression, however, had only temporary success in putting down the peasant struggles which soon re-emerged in a more militant manner.

### *Armed Revolt and the Malabar Tenancy Act Amendment*

The Congress government, even after the transfer of power in 1947, failed to ameliorate the condition of the peasantry, and was reluctant to release from jail many of the leftist freedom fighters. The *janmis* on their part refused to make their waste lands available for cultivation, forcing N.S.

---

activist and participant in the Kayyur episode (V.V. Kunhambu's *Kayyur Samaracharitram*); a well-documented historical work in English (K.K.N. Kurup's *The Kayyur Riot*); and a recent Malayalam movie (*Meenamasathile Sooryan*, directed by Lenin Rajendran).

Arunachalam, the District Magistrate of Malabar, to work out a *punam-charter*, as reported by him in January 1948:

> According to the revenue practice of the district, large extents of hilly and jungle areas are noted as unoccupied dry land which technically belongs to the landlord who is not liable to pay any assessment in the year when it is not cultivated. This freedom from assessment has made the landlords not very anxious to allow punam cultivation especially as complications regarding fixity of tenure when cultivation is permitted is likely to crop up . . . . During my tour, . . . I succeeded in assembling all the important Janmies of north Malabar and a number of ryots and evolved an understanding regarding punam cultivation which is most reasonable and satisfactory to the cultivators. The rent for punam has been fixed at a low figure and denial of land for cultivation on any ground including existence of past arrears has been agreed to be stopped. I am taking steps to give due publicity and effect to this arrangement which is more or less a *Punam Charter* to the area in question . . . . By these means I am taking steps to remove legitimate agrarian grievances which were being exploited by the communists.[31]

As nothing much came out of this charter, the infuriated peasantry organised 'paddy purchasing' from principal *janmis* with a view to preventing hoarding and blackmarketing. Batches of peasant and communist volunteers armed with available weapons went from place to place for 'forced-purchase' of paddy at a fair price though many recalcitrant *janmis* registered their protest by not accepting any payment. The paddy so taken were distributed among the poor villagers at a fair price. Slogans like 'Telangana way, our way' (inspired by the Telangana People's Armed Struggles); and 'land to the tiller and power to the people' (the Calcutta thesis of the Communist Party), lent these groups the necessary moral courage in their venture.[32] Police attacks on these groups in

[31] See GO 630 Confidential (Public) General Administration Department, 12 March, 1948.

[32] Newspapers reported these happenings in great detail. See, for instance, the *Mathrubhumi* of April-May 1948.

different places such as Korom, Thillenkeri, Peralam, Onjiyam and Aralam resulted in the death of at least twenty-two persons during April-May 1948, and the injury and arrest of many.[33] While the local newspapers characterised these happenings as 'Malabar Communist Rebellion,' in a secret dispatch the Inspector General of Police C.K. Vijayaraghavan reported that they more than confirmed the plans for an armed revolt as indicated in March 1948 and that there was ample justitication for declaring the Communist Party unlawful in Malabar.[34] The 'revolt' was, however, quelled by pressing the army into action.[35] The attacks launched by the Congress Party, the *janmis* and the police even after the 'revolt' was quelled, solely with the object of liquidating the communist movement, resulted in many more deaths of peasant and communist activists in different parts of Malabar. Particularly notable in this connection is the death of twenty-two persons in a firing in the Salem jail on 11 February 1950, of whom nineteen had been arrested earlier from Malabar. This incident has been compared to the gruesome Jallianwalabagh tragedy by the peasant and communist leaders of Malabar.

As a result of these agitations and unrest, and based on the recommendations of the Tenancy Committee of 1940, the

[33] On peasant activism in this period, the fortnightly report for the second half of April 1948: 'Communists were fairly active during the fortnight and took every opportunity to foment strikes and general unrest and to incite mobs to violent acts mainly against the police. The situation was rather bad in Malabar, South Kanara, and Kistna districts. Orders have been issued for the detention of important communists in all districts, but many of them have gone underground. In Malabar their activity has taken the form of obstructing the movement of foodgrains, of removing foodgrains from the storehouses of important landlords and of forcibly removing licenced firearms from the houses of the licencees . . . . In South Kanara district they have been organising peasant raids and peasant marches, the mode of attack being to collect men, women and children on the promise of getting them rice and to take them to the houses of landholders who are reported to have some paddy and to take delivery of the paddy by threatening the landholder with violence.'

[34] See Government of Madras, Under Secretary's Safe, GO 71, Confidential, 29 November, 1948.

[35] The strong peasant upsurge in the post-war years was a nationwide phenomenon. For details, see Kotovsky, 1964:39, 41. For the agrarian disturbances in different parts of Madras Presidency, including Malabar, see the various fortnightly reports for 1948.

Malabar Tenancy Act of 1930 was further amended in 1951. But many of the provisions of the amendment (such as depositing of one year's rent as advance by the *verumpattam* tenants, eviction of tenants by the *janmis* on a number of grounds, and failure of the Act to protect the interests of the landless homestead tenants) made the peasants even more furious. The convening of the *Malabar Karshaka Conference* in the then Kottayam taluk in May 1952, and the subsequent conferences in the other taluks, were indications of the renewed vigour which the struggles had gained during this period. Though another amendment to the Act was introduced in the Assembly in 1952, it was not different from the previous one in any significant respect. Therefore, the *Malabar Kudiyan Special Conference* organised at Calicut in December 1953 by the *Kisan Sangham*, presided over by the *Sangham* president K.P.R Gopalan and participated by more than 400 delegates and 15 'fraternal delegates,' gave a call to all the *Kisan Sanghams* to observe 15 January 1953 as 'token Satyagraha Day,' protesting against the policy of the government. It also resolved to take out a *kisan jatha* to Madras and present a mass-petition with a lakh of signatures when the Tenancy Amendment Bill came up for discussion in the Assembly (*Mathrubhumi*, 23 December 1953).

This campaign had some effect. The Malabar Tenancy (Amendment) Act of 1954 contained provisions for the fixation of fair rent, qualified the fixity of tenure, curbing the *janmi*'s right to eviction, protecting the rights of homestead tenants, and abolition of the practice of depositing one year's rent as advance by the *verumpattam* tenants (see, Varghese, 1970:42–43).

The introduction of this Act was followed by the merger of Malabar with the Travancore-Cochin state in November 1956 to form the unified Kerala State, marking the culmination of another ongoing democratic movement for state formation. With this the history of peasant struggles and land reforms in Malabar entered a new phase, which will be discussed in the following chapter.

## APPENDIX A

Petition purporting to be addressed by certain Mussulmans, Nayars, Tiyyans, and men of other castes inhabiting Malabar.
States,

1. that as a terrible outbreak is going to occur immediately in Malabar, urgent orders are requested to be issued with a view to prevent any lawlessness.
2. That since the British acquisition of Malabar, no European officers deputed to the district have received any bribes from the natives or otherwise unjustly acted.
3. That the native officials of the district take bribes and oppress the ryots; that though this is difficult to prove, on proof of any instance, the party concerned gets punishment at the hands of his European superiors.
4. That during the late famine your Grace protected the lives of innumerable people at an expense of countless sums of money.
5. That under other Governments, petitioners hear that the subject-people suffer from great oppression.
6. That the ryots are aware that such a paternal Government as this exists nowhere else.
7. That the severity of the oppression of the Malabar landlords will lead to great disturbances, at which a great number of people will have to lose their lives. This is a matter of great regret.
8. That the grievances of the subjects are stated below. The property of a great part of the lands in Malabar belongs to Hindu temples and to Nampuderies and others who are Hindus.
9. That when such lands lay waste, the Mussulmans, Nayars, and other castes obtained grants on Kanam and Otti rights, made improvements and regularly paid, besides the annual rents, the renewal fees at every twelve years at the rate of 20 per cent. of the principal sum advanced upon the original grant; that these fees are paid as mere gifts, and not secured in the vouchers obtained; that these ryots also pay the Government revenue and enjoy the surplus profits.
10. That when one has made the necessary improvements upon a grant so obtained, and when the land and improvements have reached a period when a return in the shape of profits may be expected, the grantee becomes old and dies leaving behind a large family whose maintenance is to be derived from such property, when the grantor appears, brings a suit, makes out that the improvements or less than a tenth of their real worth, gains over the Munsifs, deducts the value of improvements for Court costs, and ejects a respectable family. Thus reduced to beggary, houseless, and unable to feed a number of hungry children, one will not, of course, hesitate to give up his life.
11. That ejectments were rare in Malabar. Any hard-hearted landlord who resorted to such extreme measures would soon be killed, and a disturbance would ensue, to quell which a military force had to be

brought, and much money expended; that within the last two years the landlords, having bought over the District Munsifs, have commenced to oust the lessees, and whatever they gain is divided between themselves and the Munsifs.

12. That such a proceeding serves to enrich the landlords and Munsifs, but reduces the well-to-do lessees of land to beggary.
13. That the Mussulmans are the people committing riots: all the Hindu officials and landlords impress this fact upon the European officers of the district; that there are no Mussulmans holding high offices or acquainted with English; that this accounts for the Mussulmans being declared the principal offenders.
14. But that the fact of Mussulmans becoming offenders and losing their lives is due to their being oppressed by landlords and the Nayar officials who side with them.
15. That demolition of mosques, religious persecution, cruel oppression, and ejectment of Mussulmans by landlords are the causes that have led to Moplah outbreaks in Malabar.
16. That the assassination of Collector Conolly was committed by some Moplahs of bad character whose continued imprisonment made them despair of being liberated from the jail.
17. That the former rulers of the country and their adherents, the landlords, feel extremely dissatisfied with the subjugation of the country by the British government; that the ryots feel satisfied when they compare the welfare and tranquility they enjoy under the British rule with the oppression they suffered before; that this vexes the late rulers and landlords who therefore cause and foster dissatisfaction and riots.
18. That the Travancore Government having been fully persuaded of the fact of the holders of land being unjustly turned out and also persecuted for their religion by their landlords, have prohibited the dispossession of lessees, who are, however, declared liable to pay the annual rent and the renewal fees.
19. That no riots of the kind occur in that Native State; that within the last two years all the landlords in Malabar, assisted by a few Munsifs, have combined in filing ejectment suits and in obtaining and executing decrees in such suits.
20. That the people having, therefore, conspired to create a disturbance are advised by some wise men to wait until a representation of the popular grievances has been made to Government and orders received thereupon.
21. That whatever enactments may be passed, before all such suits have been decided and all such decrees executed, disturbances and bloodshed of a kind unknown in Malabar will take place; and that this should not be construed into a vain threat held out to deceive. By the Almighty God who has created all, petitioners swear that this will be a fact.
22. What the Government ought to consider is whether it will be proper, for the sake of enabling the landlords and Munsifs who are

comparatively few to amass immense wealth, to subject a great many people to cruel oppression, and thereby allow countless lives to be lost.

23. Petitioners pray that orders may issue prohibiting the trial of such suits and the execution of such decrees, and forbidding the registration of documents effecting transfers of lands recovered by such suits; and petitioners also request the deputation of an efficient and impartial Commissioner to inquire into complaints against landlords. The expenses of the Commission, petitioners add, may be defrayed by Court-fee stamps of one anna being required to be affixed to petitions that may be presented to him.

(No names or signatures)
(A true translation)
14 October 1880 (Signed) E. Thompson
Malayalam Translator to Government.

**Source:** Reproduced from Government of Madras, 1885:14–16.

## APPENDIX B

According to the statutory tenure recommended by Logan, the actual cultivator of a holding. Being the property of a single *janmi*, and not exceeding in extent 25 acres of wet/dry grain crop land, or five acres of garden land, was to have the following rights:

1. To permanency of tenure with remainder to his heirs, representatives, or assigns.
2. To utilise to the best advantage for agricultural purposes the soil of his holding.
3. To sell, give, or transfer his interest in the whole of his holding, and to sell (with *janmi's* written consent) his interest in a portion of his holding at the best price he could obtain.
4. To a share not less than one-third—his ancient customary share—of the average annual net produce of his holding, estimated in kind, not in money, at the time of his entry on possession.
5. To the government share of the produce estimated in kind and converted into money at the commutation rate fixed by the government, if the land revenue assessment was payable by him.

The landlord of a holding, not exceeding 25 acres of wet/dry grain crop land or 5 acres of garden land, was to have the following rights:

1. To reserve (subject to clause 3 above) as rent payable periodically, either in kind or in money, a share not exceeding the remaining two-

thirds of the average annual net produce of the holding, estimated in kind, not in money, at the time of entry on possession by the actual cultivator.

2. To sell to the actual cultivator on entry into possession, at the best price he could obtain, the actual cultivator's right to raise improvements on the holding and the actual cultivator's right to his ancient customary share, one third, of the net produce.
3. To preempt at sales, whether public or private, the actual cultivator's interest in his holding.
4. To *veto* (a reasonable right) against the person to whom the actual cultivator may wish to transfer his interest in his holding.
5. To *veto* on the sale of a portion of a holding by the actual cultivator, and a right of *ouster* if the actual cultivator mortgaged, subdivided, or sublet his holding.
6. To recover the land from the actual cultivator for reasonable schemes for utilising the land for other purposes than agriculture, or for improving generally the cultivation of his property.
7. To recover arrears of rent (not being less than one year's arrears) by the sale of the actual cultivator's interest in his holding, and to recover any arrears as a first charge on the proceeds of any sale, public or private, of the actual cultivator's interest in his holding.
8. To let his land for temporary purposes (to prevent the land from lying waste till some contingency happened).
9. To eject the actual cultivator if he made any changes in the holding with a view to cause wilful damage to the *janmi*. (Government of Madras, 1882, Vol. 1:lxxxiii–v)

# 4

# Peasant Struggles and Land Reforms, 1957–70

The formation of the unified Kerala state helped the peasant organisations of the constituent regions consolidate their position further and mobilise the peasantry better.[1] It also provided a broad political and electoral base to the Communist Party which had for long been championing the cause of the peasantry. Though these developments created, in Kerala as a whole, certain conditions for ushering in a new era of land reforms, given India's new status as a liberal parliamentary democracy with the constitutional inviolability of private property, the fruition of such reforms depended on the interplay of several other conditions: the mobilisation of the peasantry; the counter-mobilisation of the landed interests; the trials of strength in the state legislature between political parties representing the peasantry on the one hand and the landed interests on the other; the political leadership at the Centre

[1] Although peasant struggles were relatively more strong and protracted in Malabar, during the two decades preceding the formation of Kerala they began to strike roots in Travancore and Cochin also. For details, see, Balaram, 1973; Gopalan, 1975; Menon, 1969; Nair, 1973; Namboodiripad, 1943; Oommen, 1985; Pothuval, 1976.

whose assent is a constitutional imperative for the enactment of land reforms legislations; the political interests represented by the different sections of the central and state bureaucracies; and so on. Consequently, any attempt at understanding peasant struggles and land reforms since the formation of the state till the present should necessarily be in the larger socio-political context

## Communist Ministry and the KARB: 1957–59

The Communist Party which formed the first ministry in Kerala on 5 April 1957, after the general election to the state assembly held in February-March 1957,[2] was true to its election promise of introducing progressive changes in Kerala society.[3] One of the crucial spheres of such changes was agrarian relations. Within a week after assuming office, on 11 April 1957, the ministry promulgated an Ordinance (later enacted as the Kerala Stay of Eviction Proceedings Act of 1957), staying from this date all existing eviction and allied proceedings against all categories of tenants and preventing courts from accepting fresh eviction suits.[4] Namboodiripad, then chief minister of Kerala, described this Ordinance as the first ever foolproof anti-eviction measure introduced anywhere in India, inasmuch as

> in contrast to the anti-eviction legislation of Congress governments, this particular Ordinance closed all loopholes, providing that anybody who was in occupation of a

[2] Of the total assembly seats of 126, the Communist Party secured 60, as against 43 secured by the Congress Party, 9 by the Praja Socialist Party, 8 by the Muslim League, and 6 by independents (5 with the backing of the Communist Party). The percentage of votes secured by the Communist Party was 35.3, as against 37.8 by the Congress Party, 10.8 by the PSP, 4.7 by the Muslim League, 5.5 by the Communist Party sponsored independents, and the rest by others. See Lieten, 1982:25; *Mathrubhumi*, 21 March and 6 April 1957; Nossiter, 1982:123.

[3] For a discussion of the election manifesto of the Communist Party, see Fic, 1970:68–74; Nossiter, 1982:121–22.

[4] See, for details, *Mathrubhumi*, 19 April 1957; also, Government of Kerala, 1968:25–26.

> piece of land or house-site on a particular day—regardless of the tenure on which it was being held—would get the protection of the law. (Namboodiripad, 1981:46)

This Ordinance was, however, only a prelude to the comprehensive Kerala Agrarian Relations Bill (KARB), which was under the active consideration of the ministry. On 9 November 1957, K.R. Gowri, then revenue minister of Kerala and the chief architect of its land reforms, announced the important provisions of this Bill (*Mathrubhumi*, 10 November, 1957). A few weeks later the Bill was published as an extraordinary gazette on 19 December, and introduced in the assembly on 21 December (*Mathrubhumi*, 20 and 22 December, 1957). After long and laborious consultations involving the various interest groups, the work of a select committee, and debates in the state assembly, the Bill was eventually passed by the assembly on 10 June, 1959, and sent for the approval of the President of India on 27 July 1959.[5]

As a measure intended to reform and regulate tenancy with a view to its abolition, the KARB proposed fixation of fair rent at generally between one-fourth and one-twelfth of the gross produce, cancellation of all rent arrears as on 11 April 1957 on payment of the arrears for a specified period, fixity of tenure to all categories of tenants, restoration of rights to tenants evicted after the formation of the state, vesting in the government the rights of *janmis* and intermediaries on all tenanted lands for transfer to the tenants on payment of a fixed purchase price, ceiling on agricultural holdings at between 15 and 25 standard acres, and surplus land distribution among the landless and the land-poor.[6]

[5] The select committee met on 60 days, examined 249 witnesses, and received thousands of petitions. The assembly discussed the various sections of the Bill for about 114 hours in the course of 46 days. It received notices of about 2,000 amendments, discussed 1,035, and accepted 290 of them. See *Mathrubhumi*, 26 March, 17 April and 11 June 1959, and 7 October 1960.

[6] For a detailed summary of the important provisions of the KARB, see Annexure A to this chapter. For a discussion of these provisions, see Andalat, 1978:201–26; Bergmann, 1984:52–55; Herring 1983:163–68; Lieten, 1982:82–84; *Mathrubhumi*, 20 and 22 December 1957; Nossiter, 1982:149–53; Oommen, 1985:108–10; Raj and Tharakan, 1984:44–47; Varghese, 1970: 143–45.

Like the Ordinance, the KARB was the first of its kind attempted anywhere in India. Covering as it did the three distinct regions of Travancore, Cochin and Malabar, and parts of South Kanara, each with its peculiar multi-layered and caste-class reinforced land system and history of land tenure legislations, it had to tackle tenurial problems of multiplex complexity. And, attempted as it was by a political party with radical postures, but constrained to work within a liberal constitutional framework, an elitist bureaucracy, and court system, and above all, a highly hostile and anti-communist environment of political opposition in the state, at the centre, and in the rest of the country, it was probably inevitable that the Bill should have been punctuated by moderation and caution, concessions and compromises. This was evident from the provisions for payment of compensation for the lands taken over by the government (though well below the market price), the rather high ceiling relative to the availability of land, exemption of a wide variety of lands from ceiling, especially those under plantation crops, absence of any distinction between supervisory and cultivating tenants in the provisions for fixity of tenure and right of purchase of ownership to tenants, and so on. But, seen against the programmatic of the Communist Party such punctuations were necessary conditions for implementing its land reforms objective. While introducing the KARB in the assembly, K.R. Gowri stated the objective of Kerala land reforms as the implementation of the slogan 'land to the tiller' the principal slogan of the nationalist movement, within the framework of the Indian Constitution. As the first stage of the Communist Party's programmatic towards this end, the scope of the KARB was limited to the implementation of 'the recommendations made by the Land Reforms Panel of the Planning Commission—recommendations which had been approved by the Congress Party but which every state government headed by the Congress ignored in practice' (Namboodiripad, 1981:47); or, paraphrased analytically, 'to shatter the parasitic domination of the jenmis, landlords and

[7] For a discussion of the Communist Party's approach to land reforms, see Herring, 1983:163–68; Lieten, 1982:66–67, 80; Nossiter, 1982:149–51; Raj and Tharakan, 1984:43–44.

intermediaries who, in collaboration with the usurious money-lenders, made any reasonable development of the forces of production impossible' (Lieten, 1982:80).[7] What is, however, more relevant to note here is that even construing the concessions as limitations, the KARB still held out enough hope to the potential beneficiaries as to rally them behind it, and threat to the landed interests as to range them against it.

### *The KARB and Peasant Mobilisation*

The active involvement of the peasantry in the preparation and legislation of the KARB took the forms of numerous study classes, meetings, conferences, processions from the village to state levels, and evidence before the select committee on KARB—all organised by the *Kerala Karshaka Sangham* (KKS).[8] It was so intense and inspiring that even a Congress Party leader like Malaviya praised it as 'something unique, a parallel to which can hardly be found in any other Indian state' (Malaviya, 1958:89). A brief look at some of the events will give a general idea of this mass involvement.

The KKS Working Committee meeting held at Trivandrum in May 1957, presided over by KKS President Pandalam P.R. Madhavan Pilla, discussed at length the KARB in the light of the guidelines issued by the Planning Commission. The Calicut District *Karshaka* Conference held at Calicut in October 1957, presided over by M.K. Kelu, inaugurated by C.H. Kanaran (both MLAs) and attended by many peasant and communist leaders including the *All India Kisan Sabha* (AIKS) President A.K. Gopalan, KKS Secretary K.A. Keraleeyan, and A.V. Kunhambu, MP, recommended certain amendments.

The second annual conference of the KKS held at Kaladi[9]

[8] Following the formation of the Kerala state the peasant organisations of the constituent regions came together and at its first conference held on 29–30 December 1956 at Shoranur formed the unified *Kerala Karshaka Sangham*, with Pandalam P.R. Madhavan Pilla as president, and A.V. Kunhambu as secretary. See Pothuval, 1976:128; Raja, 1974:182.

[9] It is significant to note that the 9th All India Sarvodaya Conference led by Vinoba Bhave was also held at Kaladi, the birth place of Adi-Sankara, in May 1957. About 25,000 persons including Sarvodaya leader K. Kelappan, AICC President U.N. Dhebar, Kerala Governor B. Ramakrishna Rao, and

in the same month, did the groundwork for the KARB. It was attended by A.K. Gopalan, K.R. Gowri, K.A. Keraleeyan, P.K. Kodiyan, MP, law minister V.R. Krishna Iyer, A.V. Kunhambu, Pandalam Madhavan Pilla, E.M.S. Namboodiripad, about 370 KKS delegates representing an estimated 186,000 KKS members, and 200 'fraternal delegates' from different parts of the state.

The mammoth demonstration at Trivandrum organised by the KKS on 29 March 1958—that is, on 'Kayyur Day' when the Kayyur comrades were hanged—participated by A.K. Gopalan, E.M.S. Namboodiripad, AIKS General Secretary N. Prasada Rao, and reportedly by about 12,000 volunteers from different parts of the state led in enthusiastic *jathas* by A.K. Gopalan, K.P.R. Gopalan, K.A. Keraleeyan, A.V. Kunhambu, Pandalam Madhavan Pilla, V.M. Vishnu Bharateeyan and others, strongly supported the KARB. Slogans hailing the communist ministry and the KARB reverberated the air. Speaking on the occasion, A.K. Gopalan aired his apprehension of the attempts of the vested interests to thwart the Bill and appealed to the masses and the opposition parties to cooperate with the government. The KKS special conference held at Muvattupuzha in May 1958 discussed the KARB and suggested certain amendments. When the KARB was in the select committee stage, the KKS organised conferences of its representatives in each district, discussed the draft Bill, prepared amendments, and based on them gave evidence before the select committee. While giving evidence it also organised solidarity rallies and demonstrations in different places.

The third annual conference of the KKS held at Perintalmanna in March 1959 (also preceded by a mass rally), presided over by Pandalam Madhavan Pilla, and inaugurated by N. Prasada Rao, expressed its anxiety about the delay in legislating the KARB and demanded passing of the Bill as

---

Madras Chief Minister K. Kamaraj, were reported to have participated in it. It was followed by Bhave's *padayatra* for *bhoodan-gramdan* and *sampattidan*. It is not clear whether the choice of Kaladi for the KKS conference was a mere coincidence or a well thought out attempt to neutralise the Congress-Sarvodaya effects on the Communist Party's land reforms efforts.

amended by the select committee in the next assembly session itself. Following this, E.M.S. Namboodiripad assured that he would pass the KARB if his ministry did not fall within the next one and a half months.[10] And pass he did, though as it turned out, about one and a half months after passing the Bill, and about four days after its despatch for the approval of the President, his ministry was pulled down by the combined opposition of the landed interests and the political parties representing them.

### *The KARB and Counter-Mobilisation*

The opposition of the landed interests to the KARB was clearly evident from the quick rise in the forced eviction of tenants often resulting in clashes.[11] Responding to a member of the Communist Party, Namboodiripad stated in the assembly in March 1958 that the government had received several complaints of *janmis'* evicting tenants, and that such evictions were on the increase since the introduction of the KARB. When the KARB was taken up for consideration in the assembly in April 1958, K.R. Gowri reiterated this statement by reporting that in different parts of the state there were organised efforts to evict tenants from the lands which they had been cultivating for years, that though the small land owners who feared the loss of their lands formed one section of those who resorted to evictions, the big land owners who tried to thwart the KARB by forcibly taking back as much of their lands as possible were in the forefront (*Mathrubhumi*, 15 March and 3 April 1958).

[10] For brief reports on KKS activities during 1957–59 centering round the KARB, see, *Mathrubhumi*, 30 May 1957 to 31 July 1959; Nayanar, 1982:85; Pothuval, 1976:138–44; Raja, 1974:183–85.

[11] Some of these forced evictions were reported in the local press. For instance, in March 1958 the attempts of the Ottapalam *Karshaka Sangham* to rehabilitate a tenant on the land from which he was evicted by a Nair *janmi* resulted in clashes. In the same month, at Kizhoor in Ottapalam taluk the *Karshaka Sangham* tried to hold a meeting protesting against the encroachment of a *janmi* on a lease-hold, but could not do so due to opposition. In December 1958, in Wynad, about 150 persons harvested the paddy from the fields after tying the land owner and his son to a pillar in the house, presumably protesting against the forced eviction of the tenant. See *Mathrubhumi*, 12 March, 1 April and 13 December 1958.

This opposition was mobilised and articulated mostly by the land owners associations or *Janmi Sanghams*, and partly by other associations of landed interests and by individual *janmis* themselves. The numerous *Janmi Sanghams* which sprang up in different parts of the state during the years 1957–59 and survived well beyond the communist regime, betrayed the hopelessness, ad hocism, and urgency of a panic-stricken class. They were not well-knit organisations like the KKS and its units, and lacked cohesion in regard to both organisational structure and demands. One indication of this was the arbitrary and often ambivalent use of the term 'small land owners' which varied widely among the *janmis* and their *sanghams*. Despite these limitations, they could still build up enough pressure against the KARB.

There were several reasons for this. The limitation of the *sanghams* as loose and locality specific fronts was at the same time an advantage inasmuch as they were not limited to any particular size class of land owners but covered all categories ranging from big land owners to big land owning religious institutions to small subsistence land owners. The choice of the *sanghams* of even big *janmis* to style themselves as 'small land owners associations' served to woo the really small *janmis* who formed the bulk of the land owners. The cries of the far too numerous fronts had the collective effect of clamour. The multiplicity of demands aired, varying widely among the different *sanghams*, served to protect at least the core demands. More importantly, as will be discussed shortly, in contrast to the limited legislative power of the communist ministry on the side of the KKS, in a broad sense the landed interests had the state power on their side.

The activities of these *sanghams* and other allied associations, covering meetings, conventions and conferences with occasional public demonstrations, resolutions, memoranda, petitions, press statements, deputations, and evidence before the select committee on KARB, were far too numerous to elaborate here.

Briefly, these included a memorandum to the chief minister by a member of the Chirakkal Kovilakam, and a conference of the Kerala Small Land Owners Association, in April 1957; a memorandum to the Central Coconut Committee by the Central Travancore Karshaka Samajam, and a meeting of the Alathur land owners and cultivators, in January 1958; land owners meetings in different places such as Ernad, Parali, Pattiripala, Srikrishnapuram and Valluvanad, in January-February 1958; the Valluvanad Land Owners Convention in February 1958, meetings in the same month of the land owners of the Ponnani-Tirur taluks, of Ottapalam, of the Nambutiri Yogakshema Mahasabha (more of this in Chapter 7), and of the Advocates Association of Kerala; the Ottapalam Land Owners Convention in March 1958; the North Malabar Land Owners Conference in April 1958; the Kerala Karshaka Federation Conference in November 1958; and a memorandum signed by more than thousand small land owners submitted to the chief minister and the select committee on KARB with copies to MLAs, MPs, KPCC president, Communist Party secretary, etc., in December 1958.

Central to these activities were the several modifications in the KARB demanded by the *janmis*. But there were wide variations in the specific demands and in the nature of their articulation. To cite some instances:

The memorandum from the Chirakkal Kovilakam, submitted on behalf of the small land owners of Malabar, strongly defended the right of small land owners to resume all their lands for self-cultivation. According to it, small land owners were the thousands of *janmis* and *kanam* tenants in Kerala, including several widows, children and other dependents with lands not exceeding ten acres, who subsisted only on the *pattam*, or *janmi*'s share of the crop delivered by the tenants.

The conference of the Kerala Small Land Owners Association held at Ottapalam in April 1957 resolved to protect the interests of the small land owners and to solve their problems in an organised manner. It also constituted a working committee authorised to enroll more members. Its demands were for treating as small land owners all land owning families with an income not exceeding Rs. 6,000 a year from *janmam*, *kanam*, and such other land rights, for the rights of all such land owners to resume for self-cultivation that much of their lands from which they could get an income of at least Rs. 3,600 a year, and for adequate compensation for the lands taken over by the government.

The memorandum of the Central Travancore Karshaka Samajam stated that the KARB will be harmful to the growth of coconut cultivation inasmuch as the ceiling fixed was too low and included the coconut gardens.

The Alathur land owners and cultivators meeting criticised the provisions of the KARB on ceiling and conferment of ownership on tenants as unwarranted as to jeopardise food production and the life of small land owners. Its demands were for the continuation of the payment of contract rent as agreed upon by the *janmi* and tenant, right of resumption to small land owners for self-cultivation, ceiling at 30 acres of wet land or its equivalent, and compensation at the market price for the lands taken over.

The numerous meetings held in January-February 1958 in different places demanded protection of the small land owners, and resolved to organise more and more associations under the auspices of the Kerala Land Owners Association.

The Valluvanad Land Owners Convention in February 1958 covering Ottapalam and Perintalmanna taluks, presided over by a former advocate general of Madras, K. Kuttikrishna Menon, and participated by land owners from different parts of Malabar, appealed to Keralites and *samajams* outside the state to study the KARB, hold discussions, and send their views to the central and state governments. Its demands were for income-based ceiling, treating as within ceiling limit lands with a per capita income of Rs. 3,600 or less, fixation of fair rent at half the gross produce, compensation at the market price, and eviction of tenants for failure to pay the entire rent arrears within three years.

The meeting of the Yogakshema Mahasabha at Kottayam in February 1958 demanded the *janmis'* right to resume for self-cultivation all lands within ceiling limit if the ceiling was at all fixed, and to compensation at the market price for the lands taken over. Its delegation met the chief minister and revenue minister and submitted memoranda. It also stated its intention to pursue its demands with the central government and to resort to other means.

The Advocates Association of Kerala deplored the KARB as 'expropriatory,' and feared that it would wipe out the existing class of land owners, create a new class in its place, reduce productivity, and adversely affect several banks and chit funds in the state. It demanded the exemption from ceiling of all plantations and lands under commercial crops. It also appointed a committee to study the KARB and submit a report so as to enable it to present a memorandum to the government.

The Ottapalam Land Owners Convention held at Srikrishnapuram in March 1958 participated by land owners from different parts of the taluk, threatened the government that land owners of all categories would boycott it if the KARB was passed by the assembly. The Convention also took out a *jatha* which carried placards and shouted slogans.

The North Malabar Land Owners Conference held at Tellicherry in April 1958 (also preceded by a *jatha*) stated that contrary to the claims of the communist ministry, the KARB was not in conformity with the recommendations of the Planning Commission, that the Bill would antagonise the different sections interested in land and jeopardise agricultural production, and that the fair rent fixed was too inadequate. Presiding over the Conference K. Kuttikrishna Menon contended that the KARB was intended to take away all the properties of *janmis* by the government on the pretext of giving permanent rights to tenants and fixing a ceiling on holdings. The demands of the Conference were for the fixation of fair rent in the whole state as stipulated in the Malabar Tenancy Act; no reduction in rent arrears; no ceiling on agricultural lands, and ceiling, if indispensable, on the basis of income and the individual; compensation at the market price; right of resumption to land owners for self-cultivation and house-building; inclusion of arecanut, coconut, and pepper lands in plantations; and a grant of three years time to the land owners for eviction of tenants and disposal of lands.

The Kerala Karshaka Federation Conference held at Ernakulam in November 1958, attended by land owners from different parts of the state,[12] passed a unanimous

[12] Participants in the Conference included representatives of the Kerala Land Owners Association, Trichur; Central Travancore Agriculturists Association, Kottayam; Palghat District Land Owners Association; North Malabar Land Owners Association, Tellicherry; Nayar Service Society Land Policy Committee, Changanasserry; Parur Karshaka Samajam, Parur;

resolution consenting to put an end to *janmi-kudiyan* (landlord-tenant) relations subject to certain conditions. These were the *janmi*'s right to resume lands within ceiling limit, lumpsum payment of compensation at the market price for lands taken over before terminating the *janmi*'s rights, and grant of time for mutual settlement of the claims by *janmis* and tenants. The conference resolved to submit a memorandum to the 15-member AICC sub-committee on land reforms, as also to the Planning Commission, the select committee on KARB, and to the different political parties.

Of the numerous land owners associations, the Kerala Land Owners Association was probably the most active and articulate. As part of its activities, its representatives met the Kerala chief minister in June 1957, and submitted a memorandum containing the Association's views on KARB and cautioning the government that the KARB should not be such as to liquidate the small land owners, or to cause them severe loss or hardship. They also met the revenue minister in July 1957 and submitted a similar memorandum. At a meeting held at Trichur in August 1957 the Association passed several resolutions against the KARB. Its representatives campaigned against the KARB in Delhi for a week in September 1957 and met the Home Minister G.B. Pant, Planning Minister G.L. Nanda, and Planning Commission Vice-Chairman V.T. Krishnamachari, and also submitted their views to the Planning Commission. Its executive committee meeting held at Trichur in December 1957, participated by representatives of more than 50 land owners associations from different parts of the state, criticised the KARB as unjust to the small owners, and as a threat to the economic security of the state. It authorised a six-member team to apprise the KPCC land reforms committee of its views on the KARB. It appealed to individuals and organisations to discuss the KARB and send their views to the government. At another meeting held at Trichur in March 1958, it again appealed to the land owners to organise public meetings

---

Meenachil Taluk Land Owners Association; Kerala Karshaka Samajam, Mattancherry; Palghat District Small Land Owners Association, Kunnisseri; and Vypil Karshaka Sangham, Nharakkal. See, *Mathrubhumi*, 17 November 1958.

in different parts of the state and to pass resolutions opposing the KARB. At a press conference held at Trichur in April 1959 its office-bearers stated that the KARB, as reported by the select committee, was harmful to all the land owners.[13]

### *The KARB and Political Opposition*

At the political level, in January 1958, the Kerala Pradesh Congress Committee (KPCC) appointed a land reforms committee to study and report on the KARB. At a meeting held at Ernakulam in the same month, participated by representatives of the various land owners associations (such as the Kerala Land Owners Association, and the Kuttanad Karshaka Sangham), it received memoranda and petitions from different persons, and planned a systematic attack on the KARB. It appointed another committee to report on the KARB in the light of the discussions and representations received. A joint meeting of the KPCC and the Congress Legislative Committee held at Alwaye in February 1958 passed resolutions demanding income-based ceiling, one year's time after the fixation of ceiling to sell or transfer surplus lands, compensation at the market price, etc. (*Mathrubhumi*, 4 January and 22 February 1958). A meeting of the state committees of the Praja Socialist Party (PSP) and its peasant organisation, Kisan Panchayat, held in the same month, participated by the PSP and Kisan Panchayat volunteers from different parts of the state discussed the KARB. A joint meeting of the PSP-Kisan Panchayat Committees held in the following month, presided over by PSP leader Pattam Thanu Pilla, suggested certain amendments to the KARB (*Mathrubhumi*, 7 February and 10 March 1958). In the same month, a special committee appointed by the Nayar Service Society (NSS), the socio-cultural and political organisation of the Nayars, for suggesting amendments to the KARB submitted a memorandum to the government. The memorandum suggested that the KARB should be prepared in such a way as to be generally acceptable to the different sections of the population. It recommended income-based

[13] For reports on the activities of *janmis*, *Janmi Sanghams* and other associations against the KARB, see *Mathrubhumi*, 30 April 1957 to 9 April 1959.

ceiling taking into account the money necessary for a 'decent' living, including education of children and expenditure on their future, right to resume certain types of lands within the ceiling limit, right to dispose of the surplus lands within one year after the fixation of ceiling, and fixation of fair rent based on total income (*Mathrubhumi*, 26 March 1958).

Within the assembly, participating in the debate on the Governor's address on 27 April 1957, on the first day of the first session, Congress Party member George Joseph questioned the need for promulgating the Ordinance when the assembly itself was about to meet, and claimed that it would not have served any purpose other than to help the communists encroach upon the lands of the small cultivators. PSP member P.S. Nataraja Pilla also criticised the Ordinance as unwarranted and undemocratic (*Mathrubhumi*, 29 April and 1 May 1957). When the Bill to replace the Ordinance was passed on 8 May 1957, the entire opposition walked out, the Congress Party led by P.T. Chacko, PSP led by Pattam Thanu Pilla, and Muslim League led by C.H. Muhammed Koya (*Mathrubhumi*, 9 May 1957). When the KARB was taken up for debate on 17 April 1959, after the report of the select committee on it was introduced on 16 April, there was stiff opposition from these parties. Participating in the debate on 17 April, PSP member Joseph Chazhikkadan compared the KARB to Pandora's Box, the revenue minister in charge of the Bill to Pandora and the provisions of the Bill to leprosy, tuberculosis, rabies, a scorpion, snake, wolf, and so on (*Mathrubhumi*, 18 April 1959).

While this comparison itself clearly indicated the virulence of the opposition, with an assembly dominated by determined communists and an organised peasantry crying out for radical land reforms, virulence alone would not have brought victory to the vested interests. That explains why, when the KARB was eventually passed by the assembly, even the opposition members from the Congress Party and the PSP could not help but congratulate Gowri (*Mathrubhumi*, 11 June 1959). That also explains why the opposition parties realised the futility of attempting to thwart any of the progressive policies initiated by the communist ministry through constitutional means alone rather than trying to oust the ministry itself through well-orchestrated extra-constitutional anti-communist agitations.

The opposition to the communist rule and the cry of lawlessness right from the time the ministry assumed office, the increasing confrontations and clashes,[14] the Kattampalli agitation, and so on, were all small beginnings of this war-cloud. So also the declaration of the KPCC President K.M. Damodara Menon in February 1958 to end the 'violent' communist rule in a 'non-violent' way, and the resolution of the Kerala Pradesh Congress Conference held in April 1958 to end the 'communist threat' to democracy and national progress (*Mathrubhumi*, 7 February and 26 April 1958).[15] Of these, the Kattampalli agitation itself clearly showed how the political parties representing the landed interests made use of even the hapless Harijans in furthering their cause. The agitation, from 17 October to 4 December 1957, was for the rehabilitation of nine Harijan families evicted by the government from the government land occupied by them at Kattampalli (about 5 miles from Cannanore) at the instigation of the Congress Party and the PSP. The joint agitation organised by these parties involving protest marches, meetings, hartals, and satyagraha resulted in several arrests, and served its purpose of embarrassing the communist ministry at least for the time being. The agitation was withdrawn on 4 December 1957 following government's decision favouring the Harijans in the distribution of government waste lands.[16]

[14] To cite some instances of these clashes: the attempts of the anti-communists to ridicule Punnapra-Vayalar martyrs' day on 27 October 1957 as 'betrayal day' by taking out processions resulted in clashes at Vayalar and some other places in Alwaye. In November and December 1957 there were clashes between members of the KKS and others in Neyyattinkara taluk. In September 1958 in an encounter between two groups in a land dispute in Kasargod two persons were killed and some injured. Protesting against this the *Kasargod Taluk Karshaka Sangham* held a meeting presided over by K. Madhavan. After direct enquiry A.K. Gopalan blamed the Congress Party and the PSP for fomenting trouble. See *Mathrubhumi*, 28 October and 5 December 1957; 13 December 1958

[15] The Conference held at Cannanore was inaugurated by Congress Party President U.N. Dhebar, and attended by Jawaharlal Nehru, Indira Gandhi, V.K. Krishna Menon, Mysore Chief Minister S. Nijalingappa, and delegates from different parts of Kerala. See *Mathrubhumi*, 26 April 1958.

[16] For details of the agitation, see *Mathrubhumi*, from 18 October to 4 December 1957.

It was, however, only towards the close of May 1959, when the KARB was about to be passed by the assembly, that the war broke out. That was when Mannath Padmanabhan, leader of the NSS, the self-appointed custodian of Kerala's culture and society, and saviour of the anti-communists, gave a call for a joint agitation of the different sections of society irrespective of caste and creed to end the communist rule, and characterised the long lurking opposition to the ministry's various policies[17] as mere undercurrents of this impending upsurge. He also gave a call for closure of all educational institutions so long as the communist ministry's educational policy was in existence, and for the formation of committees and volunteer senas for people's support. He was speaking at Calicut under the auspices of the All-Malabar Private School Managers Association (*Mathrubhumi*, 27 May 1959). Within a week after his threat, the Congress Party tried to unite all those who were disgruntled with the communist ministry in one way or another. As Namboodiripad wrote, they were the *janmis* 'who were deprived of their former privileges; the planters and industrialists who were unable to get the police at their beck and call when there was a strike; the managers of privately-owned educational institutions on whom some control was being established and 'sections of the people who were, in fact, benefited by the land reform and education bills as well as by the police policy of the government, but in whom the poison of anti-Communism had been injected' (Namboodiripad, 1981:49–50). Accordingly, in league with

[17] Apart from land reforms, the communist ministry's police and educational policies were highly controversial. According to its police policy, the function of the police was 'tracking down and bringing to book antisocial elements who indulged in theft, cheating, murder and other crimes, rather than suppressing the fighting peasants and workers'. With the declaration of this policy employers and *janmis* 'found that they could not depend on the police machinery to suppress those working under them as they could do before.' The result was a new fillip to the tempo of the mass movement in the state (Namboodiripad, 1981:48). The educational policy of the ministry provided for the better organisation of general education, especially the private management of educational institutions, and more specifically for a check on the payment of salaries to and service security of the teachers. There was stiff opposition to the Bill from the managements of private schools, especially from the Christian churches which have a virtual monopoly of private educational institutions in Kerala. For details, see Lieten, 1982: 33–49; Nossiter, 1982:153–57.

the NSS, Christian churches, the PSP and the Muslim League, the Congress Party took upon itself the leadership of what has come to be known as the 'liberation struggle' (*vimochana samaram*) with the declared aim of saving the state from communist attacks and establishing peace, democracy, and democratic government. It formed a joint steering committee (Liberation Struggle Samiti) of the Congress Party, PSP and the Muslim League, with the KPCC President R. Sankar as president, and P.T. Chacko, Panampilli Govinda Menon, P.P. Ummerkoya, C.M. Stephen, and others as members. At the instigation of this committee 12 June 1959 was observed as 'liberation day' with statewide hartals and public meetings demanding the immediate resignation of the communist ministry. With this began the 50-day long statewide agitation of picketings of government offices and educational institutions, and closure of schools, culminating in the dismissal of the ministry on 31 July 1959 and the imposition of President's rule on Kerala for the next seven months.

The dismissal of the ministry was mainly an outcome of the Congress Party's intolerance of the communists and indignation to the communist rule. Some indications of this are available even in the Kerala Governor's report to the President of India stating that under the prevailing conditions it was impossible for the ministry to continue in power and recommending re-election after dismissal of the ministry and the imposition of President's rule on the state. In keeping with the demands of the opposition parties, the Governor reported that under the communist rule life and property were unsafe, that there were administrative discriminations, attempts at instilling communist dogmas in students, and neutralising government employees, especially the police, and that the education bill meant to regulate private educational institutions was seen as too extreme and as an attempt to teach communist dogmas (see, *Mathrubhumi*, 18 August 1959).

Insofar as the KARB was a major contributory factor to the fall of the ministry, it has to be seen in the context of the strength and weaknesses of the mobilisation of the peasantry and the counter-mobilisation of the landed interests. Though the Communist Party made tremendous efforts to mobilise the peasants and workers and to protect and promote their interests, its influence might have been limited to the 40.8

per cent votes secured by it, whereas the landed interests could draw on their side the vast bulk of the remaining non-communist electorate by arousing revivalist, socio-cultural and, above all, anti-communist sentiments. Further, the communist ministry had only limited legislative powers, whereas the landed interests had the might of the larger state power dispensed specially through the Congress Party and the judiciary. According to Namboodiripad, the provisions of the KARB were so much in the interests of even the bulk of those who rallied under the anti-communist banner (Namboodiripad, 1967:226). If this was so, it remains a moot issue if his ministry could not have had a more pragmatic programmatic to bring into its fold this bulk of the population. So also the vast bulk of the small land owners who had nothing much to lose but their anti-communist prejudices and apprehensions.[18]

The fall of the ministry was undoubtedly the loss of the peasantry. This loss was, however, far outweighed by their overall gain from the communist regime. For, as Gough wrote, 'the communist period so deepened the existing class struggles and liberated propertyless people from the old ties of subordination, that traditional relations of private property, tenancy, and wage labour (could not) be made to work harmoniously' (Gough, 1978:767). More importantly, as asserted by Namboodiripad himself, the KARB was the biggest achievement of his ministry's turbulent existence of just 28 months (Namboodiripad, 1981:47). In subsequent years neither the landed interests nor any ministry could undo the leap forward in mass consciousness which his ministry effected through the mass involvement of the peasantry in the legislation of the KARB (Lieten, 1982:84). Far from succumbing to pressures, the KARB brought to the fore land reforms as a crucial factor in Kerala's politics, determined the nature and direction of the agrarian changes in the state, and within a few years succeeded in getting them enacted and implemented by the communists themselves.

The continuing importance of the KARB was evident soon enough. In the mid-term poll to the state assembly held on 1 February 1960, land reforms figured as a vital issue in the

[18] For reports on the liberation struggle, see, *Mathrubhumi*, 27 May to 1 August 1959. Also, Fic, 1970:96–116; Gopalan, 1976:232–34; Lieten, 1982: 116–57; Namboodiripad, 1981:47–51.

campaign of all the leading political parties. The Communist Party's slogan was to implement the KARB and to bring the consequent new owners and those cultivating less than the ceiling area under voluntary cooperativisation (though one may wonder if cooperativisation could ever be a popular slogan anywhere in India because of the continuing obsession with private property and possession). The KPCC's slogan was to implement the agrarian policy of the Congress Party, covering provisions such as fixity of tenure, fixation of fair rent, ceiling based on income from land and its quality, protection of small land owners, and reasonable compensation to those losing land through land reforms.[19] The PSP's slogan was to legislate a fresh agrarian relations bill incorporating more 'progressive' provisions by deleting the 'unconstitutional' provisions of the KARB (see *Mathrubhumi*, 9 October and 11, 18 December 1959 and 5 February 1960).

## Congress-PSP Coalition and Dilution of the KARB

The Congress-dominated Congress-PSP coalition supported by the Muslim League which formed the second ministry in Kerala on 22 February 1960, after the mid-term poll,[20] with PSP leader Pattam Thanu Pilla as chief minister, was determined to undo the legacy of the communist ministry. Its task was very difficult to begin with, especially in the field of agrarian relations. For one thing, scuttling the KARB would

[19] The KPCC's election campaign was inaugurated by the Congress Party President Indira Gandhi at Kottayam on 7 December 1959, with a joint call of the 'democratic parties' (viz. Congress, PSP, and Muslim League) for wiping out their common enemy, the Communist Party. See *Mathrubhumi*, 8 December 1959.

[20] Of the total assembly seats of 126, the Congress Party secured 63 as against 20 secured by the PSP, 11 by the Muslim League, 29 by the Communist Party (including three independents), and one each by the RSP, independent, and the Karnataka Prantiya Samiti (see *Mathrubhumi*, 5 February 1960). The percentage of votes secured was 35.5 by the Congress Party, 14.3 by the PSP, 4.9 by the Muslim League, 43 by the Communist Party (including 6.2 by independents) and the rest by others. In terms of the votes secured the Communist Party had improved its position. As against 40.8 per cent (including 5.5 per cent by independents) in 1957, it had secured 43 per cent in 1960 (see, Lieten, 1982:160).

have incurred the wrath of the peasantry. That would have been a tactical blunder, especially because of the Congress Party's rhetoric of land reforms. For another, right from the time the ministry assumed office there was stiff opposition from the peasantry to any change in the KARB. That explains why soon after assuming office its revenue minister K. Chandrasekharan (also president of the PSP's Kisan Panchayat) tried to allay the fears of the peasantry by declaring his ministry's intention to enforce the KARB with 'favourable changes' fully protecting the interests of the peasantry (*Mathrubhumi*, 24 February 1960).

### *Peasant Mobilisation for Protection of the KARB*

In fact, the rise to power of the Congress-PSP coalition itself was ominous to the Communist Party, the KKS and the peasantry. For the partners of this coalition had already come out in their true colours in their attacks on the KARB and other progressive legislations initiated by the communist ministry. What is more, one of the important items in the election manifesto of the dominant partner of the coalition was a promise to implement the agrarian policy of the Congress Party. Under the circumstances, the Communist Party and the KKS again made all-out efforts to mobilise the peasantry through numerous meetings, study classes, petitions and processions in different parts of the state.

Thus, presiding over the KKS Working Committee meeting at Trichur in June 1960, K.R. Gowri warned that under pressures from the NSS and the *Janmi Sangham* the coalition was attempting to change the KARB. The meeting resolved to counter any such attempt. It also resolved to observe 10 June—the 'birthday of KARB'—as 'Agrarian Relations Bill Day' in the whole of Kerala with public meetings and demonstrations, and to organise a *Karshaka Jatha* led by AIKS President A.K. Gopalan (also present at the meeting) from Kasargod to Trivandrum to urge the government to enforce the KARB in its original form and spread the message of the KKS to the peasant masses to work unitedly towards this end.

The *Karshaka Jatha*, consisting of 29 peasant and communist activists (including C.H. Kanaran and KKS President

Pandalam Madhavan Pilla), began its long march from Kasargod on 29 June and reached Trivandrum on 24 July 1960, covering about 420 miles. Until the last day of the *jatha* A.K. Gopalan reportedly addressed about 10 lakh people in about 266 public meetings. In one such meeting at Calicut, he warned the government that if the KARB was amended its implementation would be with guns and lathis. Between 23 and 25 July—the last three days of the *jatha*—KKS units organised public meetings, collected signatures on postcards to send to the government, and appealed to members of the different political parties to send letters and telegrams to the chief minister in support of the Bill. When the *jatha* reached Trivandrum it was followed by solidarity *jathas* from different parts of the district. At a public meeting at Trivandrum, Gopalan and other peasant and communist leaders (including M.N. Govindan Nair, K.R. Gowri, and E.M.S. Namboodiripad) explained and emphasised the message of the *jatha*. Gopalan stated that from the enthusiastic receptions arranged by the villagers throughout the journey he was convinced that all sections of the peasantry were keen to have the KARB implemented in its original form. On 25 July he submitted a memorandum to the government. KKS units throughout the state observed this day as the 'Kerala Agrarian Relations Bill Memorandum Day'.

In a three-day meeting held at Trichur in August 1960, shortly after the KARB was returned by the President refusing assent and recommending modifications,[21] the Communist Party's State Council called upon all its organs and sympathisers, and friends of the peasantry belonging to the different political parties to organise strong mass agitations against incorporating the recommended modifications. In a long resolution the Council announced that these modifications, intended to augment the fortunes of plantation owners and big *janmis*, deprived the tenants and *kudikidappukar* of their prevailing rights, and the landless poor of the surplus lands.

The KKS delegates' special conference held at Trichur in September 1960, presided over by KKS President K.R. Gowri, discussed at length the recommended modifications

[21] For a summary of the modifications recommended by the President, see, Appendix B to this chapter.

and the course of action against them. The conference urged the peasants and their sympathisers and supporters to immediately organise strong agitations protesting against the amendment of the KARB. It also called for organising mass demonstrations at the taluk and district headquarters on 5 October, and to send memoranda to district authorities urging the government to pass the KARB without any change. A working committee was authorised to organise the programme. The Conference was followed by a mass demonstration in the town.[22]

### *Mobilisation of Janmis for Dilution of the KARB*

Countering the efforts of the KKS were those of the landed interests pressing forth their earlier demands. Delegations of the Kerala Land Owners Association and the Kerala Trustees Convention campaigned against the KARB in Delhi in September 1959 and met home minister G.B. Pant, commerce and industries minister Lal Bahadur Shastri, and the vice-chairman of the Planning Commission. The memorandum submitted to the President of India by the Kerala Land Owners Association requesting him to reject the KARB stated that in the preparation of the KARB the communist ministry had violated the recommendations of the Planning Commission, and that the immediate consequence of the enforcement of the Bill would be the dispossession of the *janmis* in Malabar and Cochin. It demanded the *janmis'* right to resume leased out lands in accordance with the recommendations of the Planning Commission, income-based ceiling, and payment of compensation as amounts on which the *janmis* could get interest equivalent to the fair rent (*Mathrubhumi*, 20 September 1959).

Delegations of both these associations met the Kerala Governor in October 1959 and submitted memoranda. The memorandum of the Kerala Trustees Convention stated that there were as many as 12,000 to 14,000 temples in Kerala and about 50,000 families depending on them for their living; and

[22] For reports on the activities of the KKS for protecting the KARB as passed by the communist ministry, see Andalat, 1978:228–29; Gopalan, 1976:248–53; *Mathrubhumi*, 3 June to 1 October 1960; Nayanar, 1982: 114–15.

that if the temples came under the ceiling provisions, the resulting fall in income would lead to the closure of many of them. Its demand was, therefore, for the exclusion of temples and Devaswoms from the purview of the KARB. In the same month, when the minister of state for home affairs, B.N. Datar, visited Kerala, a number of *janmis* submitted petitions and memoranda to him at several places, both individually and collectively, and secured an assurance from him that while deciding on the KARB interests of the land owners will be fully considered (*Mathrubhumi*, 7, 27–28 October, and 2 November 1959).

### *KARB Diluted as the KARA*

If the continuing pressures of the landed interests provided an excuse to the coalition ministry for diluting the KARB, the refusal of the President to give assent to the KARB and its return to the Kerala Governor recommending several important modifications gave it the much needed legitimacy and support.

The return of the Bill by the President was towards the end of July 1960, exactly a year after it was sent to him by the communist ministry. The coalition was quick to grab the opportunity. The legislation of a revised version of the KARB on 15 October 1960 placating and protecting a wide range of landed interests, the President's assent to it on 21 January 1961, its publication in the state gazette on 3 February 1961, and enforcement from 15 February 1961 as the Kerala Agrarian Relations Act (KARA) of 1960, all followed in quick succession (see *Mathrubhumi*, 16 October 1960, 25 January and 16 February 1961).

Though the revised Act retained the main body of the KARB due to pressures from the peasantry, even the tinkering made by the ministry had several implications for the peasantry. The deletion of the provision for restoration of rights to tenants evicted after 1956 and inclusion of the provision for eviction of *kudikidappukar* within plantation areas, deprived such tenants of their rights and legitimised the numerous evictions already effected by the *janmis* anticipating the KARB. Redefinition of the small owner as one with rights in less than ten acres of double-crop wet land but possessing only less than 5 acres benefited more owners at the expense of tenants. Further concessions to plantations (already excluded from the ceiling) by broadening the definition of plantations (so as to include in addition to plantations proper all the contiguous and interspersed agricultural lands within the plantation boundaries) and deletion of the provision invalidating land transfers after 18 December 1957 reduced the availability of surplus lands. The one year time granted to the *janmis* for resumption of tenanted lands also cut into the rights of the peasantry. Finally, deletion of the provision for elected representation on the Land Board and Land Tribunals reduced the possibility of effective implementation of the Act in the absence of representatives of the peasantry on the implementation machinery.[23]

### *Poor Implementation of the KARA*

Though the revised Act was enforced from 15 February 1961, the sections enforced from this date were only for the constitution of Land Board and Land Tribunals, fixity of tenure, and fixation of fair rent. What remained to be enforced were sections relating to compulsory take over of the rights of *janmis* and intermediaries, conferring such rights on tenants, fixation of ceiling, take over of surplus lands, payment of compensation for the lands taken over, and governmental supervision of such lands until their redistribution. These sections could have been enforced only after fixation of fair rent and the one year time allowed to *janmis* for resumption of lands (*Mathrubhumi*, 4 December 1961).

But the implementation of even the few sections already enforced was too tardy as to be of any consequence, as may be seen from the data presented in Table 4.1 for the first eight months of implementation. Against 12.7 lakh tenant households (as estimated in 1967), the Land Tribunals had received only about 58,000 applications for the fixation of fair rent, the main plank of the Act, and of them only 2,475

[23] For the provisions of the revised Bill, see Government of Kerala, 1968:26–27; for a discussion of these provisions, see Herring, 1983:171–72; Oommen, 1985:111; Raj and Tharakan, 1984:47–48; Varghese, 1970:145–46.

**Table 4.1** ***Implementation of KARA, 1960: 15 February to 31 October 1961***

| *Section* | *Provision* | *Applications* | | % |
|---|---|---|---|---|
| | | *Received* | *Disposed* | *Disposal* |
| 7 | Resumption of lands by *varam* tenants evicted after 11 April 1957 | 28 | 5 | 17.8 |
| 10 | Resumption of lands by *janmis* | 362 | 36 | 9.9 |
| 16 | Fixation of fair rent | 58,298 | 2,475 | 4.2 |
| 18 | Relief to small *janmis* | 16,308 | 367 | 2.2 |
| 36 | Purchase of *janmis*' rights by tenants | 3,015 | 60 | 2.0 |
| 37 } 78 } | Grant of *kudikidappu* rights | 10 | 3 | 30.0 |

**Source:** Tabulated from *Mathrubhumi*, 4 December 1961.

(4 per cent) were disposed of. Similarly, against 3.4 lakh *kudikidappukar* (as estimated in 1967), the Land Tribunals had received only 10 applications for the purchase of *kudikidappu*. The situation was much the same even after the first twenty months of implementation. As on 31 October 1962, the Land Tribunals had received only 79,168 applications for fixation of fair rent, and of them only 21,420 (27 per cent) were disposed of (*Mathrubhumi*, 19 February 1963). This tardy implementation was mainly because of the delay in constituting the implementation machinery, especially the Land Tribunals, and in initiating the implementation process, and the absence of any record of tenancy. It was followed by the suspension of the Act, governmental attempts to replace it with a further diluted version and mass evictions of tenants, especially in the high ranges. In these the peasantry found a systematic attack on their rights and attempts to bury deep the land reforms initiated by the communist ministry. This forced the Communist Party and the KKS to further intensify their efforts to mobilise the peasantry for protecting the favourable provisions of the KARA.

### *Peasant Mobilisation for Protection of the KARA*

As part of these mobilisation efforts the KKS formed 'Kisan Service Squads' in the villages. From 6 to 17 June 1961 A.K. Gopalan went on fast at Amaravati in order to highlight the hardships of about 1,700 families who had been evicted by the government from the forest lands of Udumbanchola taluk in Idikki and thrown into Amaravati about 40 miles away. There were also protest marches, demonstrations, and picketing of Collectorates in the state by the KKS. In September 1961 Fr Vadakkan, a Catholic priest who was active in the *Vimochana Samaram* and who, with the mass evictions in the high ranges during 1960–61 became an ally of the communists, led a petition *jatha* to the Secretariat at Trivandrum. This *jatha*—comprising *jathas* of peasants from Kottiyur, Ambalavayal, Udumbanchola and Amaravati in the high ranges—brought to the fore the problems of about 20 lakh cultivators who lived on government and private forest lands for several years. Its demands were for conferring occupancy rights on the cultivators of Kottiyur Devaswom lands, assignment of government lands to their occupants, and the grant of one acre of land to each of the Amaravati settlers. In the same month the KKS general secretary C.H. Kanaran and the Amaravati Settlers Committee president M.V. Joseph met the Food and Forest Minister Poulose and submitted a memorandum.

In October 1961 the KKS organised 'block *jathas*' in different parts of the state. Through house-to-house campaigns, discussions and meetings, these *jathas* apprised the people of the problems of the peasantry and of the proposed agitation for redressing their grievances. It also organised special conferences of oral tenants—that is, tenants without any record of rights or receipts of rent paid, and of those in the high ranges.

On 5 November 1961, after month-long preparations by its different units, the KKS organised a special conference at Trichur, participated by about 500 delegates from different parts of the state. This conference drew up a charter of demands and resolved to start an agitation if the government failed to concede to them. These demands included protection of the rights of oral tenants, full implementation of the KARA without delay by appointing more Land Tribunals, assignment of government forest lands to the peasants living on them, and fixity of tenure to occupants of private forest lands. On 13 November a delegation of the KKS met the chief minister and other ministers and submitted a memorandum.

From 27 November 1961 to 6 January 1962 the KKS organised a statewide agitation. It began with picketing of all the collectorates, but later included picketing of taluk offices, *jathas* and demonstrations, and the stormy *jatha* led by A.K. Gopalan from Kottiyur to Trivandrum for picketing the Secretariat leading to mass arrests and to the one-day general strike by the Kerala State Trade Union Congress (KSTUC). This 41-day agitation, participated by an estimated 75,000 Communist Party and KKS volunteers, had some encouraging results. At a meeting of the KKS delegates (including K.R. Gowri, C.H. Kanaran and A.K. Pothuval), and the coalition ministers held on 5 January 1962, the coalition conceded to a number of KKS demands, such as the creation of record of tenancy for tenants submitting applications to the Land Tribunals if their rights were in dispute; reduction of rent arrears from 11 April 1957 by at least half, and an increase in the number of Land Tribunals after reviewing their performance up to the end of February 1962.

When Mannath Padmanabhan warned of a statewide agitation against the KARA (more of this shortly), the KKS State Council meeting held on 10–11 July 1962 declared that it would counter it by a direct agitation for protecting the favourable provisions of the Act. It organised a special conference at Palghat on 9–10 August to work out the action-plan. As part of the preparations for this conference, 5 August was observed as 'flag-day' by the KKS when its units hoisted the KKS flags in different parts of the state. This time, however, Padmanabhan's threat remained vacuous.

When the implementation of the KARA was suspended in December 1962 following the Kerala High Court judgements in the preceding month (more of this shortly) and the government declared its intention to legislate a fresh Bill within a year, the KKS observed 30 December 1962 as 'KARA Protection Day' by organising public meetings and demonstrations in different parts of the state. When the government began its work on the new Bill, in August 1963 E.M.S. Namboodiripad reiterated that the Communist Party would not tolerate any attempt to weaken the KARA. For the next four months the agitations of the peasantry centered around the protection of the KARA and its validation by inclusion in the Ninth Schedule of the Indian Constitution.

The *Karshaka Tozhilali Party* (KTP: Agricultural Labourer Party) formed in 1962 by Fr Vadakkan played a major role in organising these agitations. The *Kerala Karshaka Convention* organised at Trichur by him and by representatives of other peasant organisations (including the KKS, inaugurated by A.K. Goplan) urged the central government not to succumb to the pressures of the state for a fresh legislation. It also gave a call for organising mass agitations of the peasantry irrespective of political affiliations. The *Karshaka Niyama Raksha Samiti* (KNRS: KARA Protection Samiti) formed at this Convention organised a statewide agitation to include the KARA in the Ninth Schedule. The agitation involved a token satyagraha by the KNRS volunteers in front of the Collectorates, a Jeep *jatha* from Kasargod to Trivandrum, and mass picketing of the Secretariat at Trivandrum. As part of this agitation, under the auspices of the Trichur District Committee of the KTP, in August 1963, about 400 persons observed a token satyagraha in front of the Trichur Collectorate. A petition submitted by this Committee to the Collector signed by about 10,000 persons demanded that the government give up its attempts for fresh legislation and include the KARA in the Ninth Schedule.

The Karshaka Convention held at Ernakulam on 3 September 1963 reiterated its demand for the inclusion of the KARA in the Ninth Schedule, and urged the central government and the Planning Commission to prevent the state from its attempts for fresh legislation. When the draft Bill—the Kerala Land Reforms Bill, 1963—was introduced in the assembly on 20 September 1963 after its publication as an extraordinary gazette on 15 September, the KNRS volunteers observed a 'black day' with mass picketings and satyagrahas throughout the state. The KNRS also submitted a memorandum to the government seeking several amendments to the Bill. When the assembly began debates on the Bill on 24 September, Communist Party MLA N.E. Balaram deplored the Bill as 'the burial of the KARA,' intended to deprive the peasantry of their rights, and to reduce them to slaves of the *janmis*. When the assembly appointed a select committee on the Bill, the Communist Party registered its protest by refusing to serve on it.

Under the joint auspices of the KKS and the predominantly Catholic *Malnad* (high ranges) *Karshaka Union*,[24] in October 1963 A.K. Gopalan, Pandalam Madhavan Pilla and others organised picketing and demonstrations in front of the Secretariat at Trivandrum, demanding the withdrawal of the Bill and inclusion of the KARA in the Ninth Schedule. On the 'Kerala Land Reforms Bill Burning Day' observed in the same month, copies of the newly drafted Bill were publicly burnt. In the following month, in black-flag demonstrations protesting against the Bill effigies of the revenue minister P.T. Chacko, and of *janmis* were also publicly burnt.[25]

Despite these protests, the peasantry again lost to the combined might of the landed interests and the government representing them. For, despite stiff opposition, the Kerala Land Reforms Bill was passed by the assembly on 4 December 1963 and by the President on 31 December. A better understanding of how the peasantry lost their battle again calls for a brief look at the machinations of the landed interests.

## *Mobilisation of Landed Interests and Dilution of the KARA*

Having succeeded in getting the KARB of the communist ministry vetoed by the President, and its revised version weakened by the government, the next strategy of the landed

[24] The *Malnad Karshaka Union* emerged during 1960–61 after Fr Vadakkan and some other Catholics began fraternising with the communists and taking up the cause of the peasantry in the high ranges (Namboodiripad, 1967:227).

[25] For reports on peasant activism during the years 1961–63, see Andalat, 1978:164–302; Gopalan, 1975:64–108; 1976:236–60; *Mathrubhumi*, November 1961 to October 1963; and Oommen, 1985:114.

interests was to thwart even this diluted version. This was a grand success. Within two weeks after the KARA was enforced began a spate of writ-petitions—3 in the Supreme Court and as many as 535 in the Kerala High Court—challenging the constitutional validity of the Act on a number of grounds, especially the violation of the rights to property, equality and freedom guaranteed in the Constitution. The extent of land reported to have been owned by some of the petitioners clearly showed that big *janmis*, who would have been the principal losers by the implementation of the Act, were in the forefront to frustrate the Act. To cite some instances—Chelur Purushottaman Nambutiri, a petitioner in the Supreme Court owned about 1,200 acres of land and was likely to lose 800 acres. Kuttipuram Kovilakam, a petitioner in the Kerala High Court owned about 70,000 acres as Devaswom property, leased out to 4,000 tenants. Kottayam Thekke Kovilakam, another petitioner owned about 1,000 acres (see, *Mathrubhumi*, 18 March and 1 August 1961).

If the stay by the courts of the implementation of the Act in the case of the petitioners until their petitions were disposed of gave the *janmis* a respite, the Supreme Court judgement in December 1961 invalidating the operation of the Act in the ryotwari taluks (Hosadurg and Kasargod) of South Kanara where the holdings did not strictly conform to the definition of an 'estate' as given in the amending Article in the Constitution restricting property rights, resulted in a resurgence of their activities.

In the first half of 1962 these activities included meetings of the Palghat Small Land Owners Association, Valluvanad Land Owners Association, and Kerala Karshaka Federation, in January; of the Trichur Small Land Owners Association, Lakkidi land owners, Ottapalam Bar Association, Cannanore District Small Land Owners Association, and the Kerala Devaswom Trustees, in March; of the Pang Small Land Owners Association, Palghat Small Land Owners Association, Karshaka Federation, and Mukundapuram Taluk Small Land Owners Association, in April; meeting of the small land owners of Srikrishnapuram, and the Small Land Owners Convention at Palghat, in May; and meetings of the Kerala Small Land Owners Association, NSS representative body, and Koduvayur land owners in June. Some of these meetings also submitted memoranda to the government and to the KPCC subcommittee on land reforms.

Central to these activities were not only the several modifications to the Act demanded earlier but also the suspension of the implementation of the Act until the incorporation of these modifications. But more often than not, these activities went well beyond the articulation of mere demands. To cite some instances: the Valluvanad Land Owners Association called on the land owners to activise the *Janmi Sanghams* everywhere, petitioning the government, and resorting to satyagraha, picketing, and revenue boycott. The Cannanore District Small Land Owners Association appealed to all land owners to write their views on the problems of the Act to the KPCC subcommittee on land reforms. The meeting of the Devaswom Trustees, attended by the Aiyappa Seva Sangham, NSS, Devaswom Confederation, and delegations of more than 50 other Devaswoms from various parts of the state, resolved to make concerted efforts to protect the Hindu religious and charitable institutions from the provisions of the Act. The Palghat Small Land Owners meeting, participated by delegates from other districts as well, warned the government of 'direct action' from 15 June 1962 if the Act was not amended and its implementation was not suspended. The Kerala Karshaka Federation threatened to launch an agitation from the second week of August. The Small Land Owners Convention at Palghat, participated by more than 200 delegates from different parts of the state (including former advocate general of Madras K. Kuttikrishna Menon and Mannath Padmanabhan), gave full backing to the call for agitation from 15 June and even constituted a 17-member action group with R. Raghava Menon as president. Inaugurating the Convention Padmanabhan warned that the Act would wipe out the small *janmis*, and strain the 'harmonious relations' between the *janmis* and tenants. He called for a strong statewide agitation against the Act, but cautioned that it may not lead to easy victory as in the case of the *Vimochana Samaram* which involved different sections of society. Later, in July 1962, he deplored the Act as a 'parting gift' from the communist ministry, and met Home Minister Lal Bahadur Shastri to seek the central government's intervention, especially for excluding Devaswom lands from the purview of the Act. The Kerala Land Owners Association passed a resolution for introducing a private bill in the assembly to protect the rights of small land owners, and constituted a committee for its preparation. It deplored the callousness of the Congress Party and the ruling Congress in regard to the demands of small land owners, and threatened to jettison the Congress by a wholesale desertion of land owners from it. The NSS representative body also warned the government that failure to protect the small land owners would have dangerous consequences. At the Koduvayur meeting R. Raghava Menon, president of the *Kerala Janmi Sangham*, deplored the KARA as 'Kerala Janmis Misery Act' and called on the *janmis* for united action against it.

While the specifics of most of the modifications demanded varied widely, central to them were the earlier demands for broadening the definition of small owners, the right of such owners to resumption, higher fair rent, and termination of tenancy for failure to pay rent, higher ceiling limit, ceiling on the basis of income and individual, grant of time for disposal of surplus lands, compensation at the market price for the surplus lands taken over, and exclusion of religious and charitable institutions from the purview of the Act.

When the clamour of the *janmis* almost reached a climax came the two judgements of the Kerala High Court, in November 1962, invalidating the operation of the KARA in Malabar and Travancore as well. Had the government been serious about land reforms it could have pursued the matter with the courts, and also got it revalidated by the seventeenth Constitutional amendment of 1964 for inclusion of certain land reforms Acts in the Ninth Schedule. Instead, following the judgements it suspended the implementation of the Act, promulgated the Karshaka-Kudikidappu Ordinance of 1962 staying eviction of tenants as defined in the Act, and announced its intention to legislate a fresh Bill. All these gave a fillip to the further efforts of the *janmis* for fortifying their rights. Their subsequent activities were focussed on preventing the possibilities of the KARA getting included in the Ninth Schedule as demanded by the peasantry, and pushing through the modifications demanded.[26]

To cite some instances. In July 1963, a meeting of the various *Janami Sanghams* held at Kottayam requested the central government not to include the KARA in the Ninth Schedule of the Constitution, and sent a delegation to Delhi to meet the concerned ministers. The Yuvajana Nayar Sahodara Samajam, Trivandrum also made a similar request. In August 1963, welcoming the government's decision to introduce a fresh Bill, the Yogakshema Mahasabha reiterated its earlier demands for income-based ceiling, the right to resume lands within ceiling limit, and compensation at the market price. A meeting of the Small Land Owners Association at Trichur requested the central government to withdraw the 17th Constitutional Amendment Bill introduced in the Lok Sabha. A meeting of the Kerala Small Land Owners Association held at Palghat, presided over by R. Raghava Menon, stated that the Bill under preparation was unacceptable to the small land owners of Kerala, especially of Malabar and Cochin. The meeting made it clear that no Bill will be acceptable to the small land owners if it denied them the right to resume land. It demanded payment of contract rent, and the immediate withdrawal of the 17th constitutional amendment. It also gave a call for 'direct action' and requested Mannath Padmanabhan to give it the necessary leadership and support.

In September 1963, a conference of the Small Land Owners Association held at Trichur resolved to organise a strong 'countrywide' campaign to apprise the government and the people of the hardships of the *janmis* and formed a committee for the purpose.

When the government introduced the fresh Bill in the assembly on 20 September 1963, there was strong opposition

[26] According to one report, the office-bearers of the Kerala Land Owners Association claimed that by this time the Association had more than 300 units and more than 20,000 members, that it spent more than Rs. 300 a month for its work. The report stated that though there was widespread sympathy for the problems of small land owners there were also widespread complaints about the leadership they had. See *Mathrubhumi*, 14 May 1963.

to it from the parties representing the landed interests including the Congress. On 23 September 1963 the PSP-Kisan Panchayat volunteers observed 'token picketing' and 'black-day' in a number of places. On 28 October 1963 about 100 persons led by the Kerala Small Land Owners Association secretary Krishnaswamy Iyer demonstrated in front of the Secretariat at Trivandrum. For a change their slogans were hymns from the Gita and Upanishads. However, betraying their class character, the demonstrators took to their heels as soon as the communist volunteers appeared on the scene for their demonstration.[27]

These protests were, however, mere pressure tactics which the government was badly in need of because of the mounting pressures of the peasantry for protecting the KARA. For, the Supreme Court and Kerala High Court judgements, the suspension of the KARA, the legislation of the Kerala Land Reforms Bill and President's assent to it in less than a month, all seemed to be exercises in aiding and abetting the claims of the *janmis*.

### *KARA Diluted as the KLRA*

The Kerala Land Reforms Act (included in the 9th Schedule) was a mutilated version of the KARB and even of the KARA. It eliminated the core provision of the earlier Bills for compulsory vesting in the government of the rights of *janmis* and intermediaries on tenanted lands for transfer to the tenants. Instead, it only provided for *janmis* and tenants working out transfer of ownership mutually through the Land Tribunals. It redefined the small owner still more loosely as one having interests in land up to 8 standard acres or 24 ordinary acres. Such owners could unilaterally resume half their holdings from those tenants who had no fixity of tenure in respect of holdings leased in immediately before 21 January 1961. Their tenants could claim ownership rights only if they agreed to surrender half the leased-in area. Alternatively, they had to pay 75 per cent of contract rent for holding the land. The Act fixed higher rates of fair rent, and ceiling at 12

[27] For reports on the activities of the landed interests against the KARA, see *Mathrubhumi*, January 1962 to October 1963.

standard acres or 15 to 36 ordinary acres for a family of five members, with the standard acre varying from 1 to 4 ordinary acres depending on the quality of land. It introduced further exemptions in ceiling, the most important of them being cashew estates with an area of ten acres or more in one block, pure pepper and pure coconut gardens of more than 5 acres, and Kuttanad *kayal* lands (paddy lands reclaimed from backwaters).[28]

While this Bill became an Act and was enforced on 1 April 1964, the ministry which enacted it soon came down. The factors responsible for its fall had begun to operate much earlier. In November 1961 the Muslim League withdrew from the coalition due to differences with its partners. In October 1962 the PSP withdrew from the coalition protesting against the appointment of Pattam Thanu Pilla as Punjab governor in the preceding month, sensing a trick of the Congress high command to weaken the position of the PSP in Kerala. To top it all the squabbles within the Congress Party came to a head in September 1964, when one section of the ruling party led by K.M. George moved a no-confidence motion. Following this the state again came under President's rule. Though a mid-term poll was held in February 1965 no party could muster a majority to form the government, and the President's rule continued till 1967 (see Fic, 1970:130–43).

### Poor Implementation of the KLRA

With the continuing political instability in the state and the low priority accorded to land reforms by the government, the chances of the implementation of the KLRA had become more bleak. This was revealed by a land reforms survey conducted by the state bureau of economics and statistics during 1966–67. According to this survey, while there was a sharp increase in the rate of creation of new tenancies and concealed tenancies (mortgages) and in the hectic land transfers for circumventing the provisions of the Act, only less than 4 per cent of the total area under tenancy in the state had been covered by the fixation of fair rent, no tenant had

[28] For a discussion of the provisions of this Act, see Bergmann, 1984:56–57; Herring, 1983: 175; Oommen, 1985:111–12; Raj and Tharakan, 1984:48–49; Varghese, 1970: 148.

obtained the ownership rights through the Tribunals, only a few had purchased ownership through mutual arrangement with the land involved accounting for only about 1 per cent of the total leased in area, and no land was acquired as surplus by the imposition of ceiling (see Government of Kerala, 1968:Chapters x, xi).

Meanwhile, a more favourable climate for land reforms was building up from April 1964, with the split in the communist movement and the emergence of the CPI(M) as an independent party.

## Leftist Coalitions and Revival of the KARB

With the rise to power on 6 March 1967, of the CPI(M)-led seven party United Front (UF), after the general election to the state assembly on 19 February 1967,[29] and after 30 months of President's rule in the state, the possibilities of reviving the long languishing KARB became a reality. For one thing, land reforms had the pride of place in the UF's election manifesto. For another, to a great extent this ministry was a replication of the first. As its chief minister and revenue minister had been the very same persons who held these offices in the first ministry and contributed much to the creation of the KARB, their task on land reforms this time was easier. What is more, there was hardly any opposition to land reforms (Menon, 1969:12).[30]

[29] Of the total assembly seats of 133, the seats secured by the UF partners were: CPM 52, CPI 19, SSP 19, RSP 6, KSP 1, KTP 2, and ML 14. Of the remaining seats the Congress Party secured 9, Kerala Congress 5, Karnataka Samiti 1, and independents 5. With the exception of the ML all other parties could be described as left-oriented. For a discussion of the 1967 election and the post-election alliances in forming the government, see Varughese, 1978.

[30] The reasons for the lack of opposition this time need to be examined. Some of them seem to have been the decade-long interval between the passing of the KARB and the present attempts at land reforms enabling the *janmis* to rearrange their holdings to a great extent; the break-up of the anti-communist alliance of 1957–60 and regrouping of the political forces since 1962, the working together of the Communist Party and some of the earlier anti-communist fronts on mass issues resulting in increasing mobilisation of the peasantry, and above all, the increasing decay and disarray in the Congress Party leading to its split in 1969.

As in 1957, the ministry first promulgated an Ordinance (enacted later as the Kerala Stay of Eviction Proceedings Act of 1967). This was on 20 May 1967 (see, *Mathrubhumi*, 21 May 1967). This was followed by a comprehensive land reforms amendment Bill, introduced in the assembly on 19 August 1968, and passed by it on 17 October 1969, as a successful attempt to restore the major provisions of the KARB of 1959.

The UF ministry resigned on 24 October 1969, just about a week after passing this Bill, due to serious differences among its partners (see, Varughese, 1978). Following this a five-party coalition consisting of half the UF partners,[31] led by the CPI with C. Achuta Menon as chief minister, assumed office on 1 November 1969. When the Bill got the President's assent on 16 December 1969,[32] it announced its intention of implementing the Act with effect from 1 January 1970 itself.

Although this coalition also fell in June 1970 due to defections, the mid-term poll held in September 1970 brought to power another coalition led by the same party,[33] with the same chief minister. This coalition continued in office until early 1977. By this time the implementation of the Act was virtually complete as will be seen in the following chapter, significantly enough, as the only land reforms Act in Kerala whose implementation has run full course, and as the only land reforms Act whose implementation has had major consequences for Kerala's agrarian society.[34]

[31] The other partners of the coalition were the ML, SSP, RSP, and the Kerala Congress.

[32] During 1971–72 this Act obtained protection of Article 31 of the Indian Constitution by inclusion in the Ninth Schedule.

[33] The other partners of this coalition were the Muslim League, SSP, RSP, PSP, Congress (R), and the Kerala Congress.

[34] There have been several other land reform enactments in Kerala since the formation of the state till the present, either to bring about parity in the land tenures in the constituent regions or to remove irritants from the principal Acts. From the point of view of this study these are of minor nature. For details of these Acts, see Government of Kerala, 1977.

## APPENDIX A

### Summary of the Important Provisions of the KARB

1. Fixation of fair rent at generally between one-fourth and one-twelfth of the gross produce, depending on the fertility of the land, and land improvements.
2. Cancellation of all rent arrears as on 11 April 1957, on payment of the rent for one year by tenants holding below 5 acres of land, three years by tenants holding 5 to 15 acres, and six years by those holding more than 15 acres, or the actual rent arrears, whichever was less.
3. Fixity of tenure to all categories of tenants including tenants at will, and occupants of homesteads (*kudikidappukar*).
4. Restoration of rights to tenants evicted after the formation of Kerala who would have normally obtained fixity of tenure under the KARB.
5. Vesting in the government on a day which it was to notify as Peasant Day the rights of *janmis* and intermediaries on all the leased out lands and enabling the tenants to purchase the same on payment of 16 times the fair rent in 16 annual instalments or 75 per cent of it in a lumpsum.
6. Ceiling on agricultural lands at 15 acres of double crop paddy land or its equivalent for a family of five members, with one additional acre for every additional member, subject to an overall ceiling of 25 acres.
7. Invalidation of sale and free transfer of lands above ceiling from 18 December 1957, with a view to preventing evasion of ceiling by potential surplus holders.
8. Acquisition of surplus lands by the government on payment of a multiple of fair rent as compensation, namely, 16 times the maximum fair rent for the first 5 acres, 14 times for the second 5 acres, 12 times for the third 5 acres, 10 times for the fourth 5 acres, 8 times for the next 30 acres, 6 times for the next 50 acres, and 5 times for the next 100 acres and above.
9. Distribution of surplus lands to the landless and the land poor such as dispossessed tenants, expropriated *janmis* with less than three acres of land, cooperative societies of agricultural labourers, and agricultural labourers.
10. Resumption of lands by *janmis* in limited cases (such as for extension of places of public worship and construction of residential buildings) after leaving the tenant at least 20 cents in his possession; and for self-cultivation if the tenant already enjoying fixity of tenure had land above the ceiling area.
11. Protection of small *janmis* (owning or holding five acres or less of double-crop paddy land or its equivalent) from the provisions granting fixity of tenure and purchase of ownership rights to tenants.
12. Exemption from the ceiling provisions of lands of religious and charitable institutions, of plantations, cashew estates, and private forests, lands housing mills, factories, and workshops, and lands not ordinarily cultivable.

13. Constitution of a Land Board at the state level and Land Tribunals under it as implementation machinery.
14. Political representation on the Land Board through a member elected by the state legislature and on Land Tribunals through members elected by local bodies.

## APPENDIX B

### Summary of the Modifications Recommended by the President of India in the KARB of 1959

1. To exclude from the definition of *kudikidappukar* those allowed to remain so on the lands of plantations, mills, factories, workshops, etc., if such persons had no *kudikidappu* rights under the prevailing laws of the state.
2. To shift *kudikidappukar* elsewhere if the land occupied by them was required for urban and industrial purposes.
3. To exclude from the definition of small owners those owning more than 10 acres or holding more than 5 acres of double-crop paddy land or its equivalent.
4. To include in the definition of plantations up to one-fifth of those other lands adjoining and within the plantations including agricultural lands which were considered necessary for the maintenance of the plantations.
5. To repeal the provision invalidating the land transfers after 18 December 1957.
6. To provide annuity to religious, charitable, and educational institutions at the rate of the fair rent received from the lands taken over from them.
7. To give priority in surplus land distribution to cooperatives formed by persons eligible for such lands.
8. To amend the provisions on the constitution of Land Board and Land Tribunals so as to specify that government will nominate persons to these bodies (*Mathrubhumi*, [illegible] August and 16 October 1960).

# 5

## Implementation of Land Reforms, 1970–81

The Kerala Land Reforms (Amendment) Act (KLRAA) of 1969, though technically an amendment to the Kerala Land Reforms Act (KLRA) of 1964, was a comprehensive legislation intended to restore the original provisions of the Kerala Agrarian Relations Bill (KARB) of the first communist ministry. With liquidation of traditional landlordism and large-scale redistribution of land ownership as its core objectives, it was undoubtedly the most drastic of all the hitherto land reforms enactments in Kerala, and probably the first of its kind anywhere in India.

The enforcement of this Act was in an atmosphere of mass agitations by its potential beneficiaries. As these agitations had a direct bearing on the implementation of the Act, a brief note on them will place the chapter in proper context.

### Mass Agitations

As a party mainly responsible for the KLRAA but out of power shortly after it was passed by the assembly, the CPI(M) rightly apprehended delay in the implementation of the Act, and made all-out efforts for building up pressure from below through its peasant organs *Kerala Karshaka Sangham* (KKS) and *Kerala State Karshaka Tozhilali Union* (Kerala State Agricultural Labourers Union). Accordingly, in December 1969 it organised a mass convention of peasants and agricultural labourers at Alleppey. The convention was preceded by a two-day delegates conference presided over by A.K. Gopalan, and participated by about 2,000 delegates including those from political parties like the KTP, KSP, and SSP, and prominent peasant and political leaders of the CPI(M) like Harekrishna Konar, E.M.S. Namboodiripad, and P. Sundarayya. Three campaign-*jathas* from the north and south of the state converged at the venue after covering several miles, and villages, and addressing numerous meetings on the way. More than three lakh persons were reported to have participated in the mass rally organised on the occasion, and about five lakh persons in the convention. The convention declared to treat the implementation of the KLRAA as a *fait accompli* through direct action from 1 January 1970. Following it, the CPI(M) organised a hectic campaign throughout the state, constituted struggle councils (*samara samitis*) at the village, taluk, and district levels, and then a statewide land-grab agitation from January to May 1970. The agitation was mainly by fencing in the *Kudikidappu* lands.[1]

Though the convention had also declared its intention of establishing the rights of cultivating tenants as owners of their leased-in lands and accelerating the acquisition and distribution of surplus lands, the agitation was focussed on the occupation of *kudikidappu* lands. Therefore, in May 1972 the CPI(M) again organised a mass convention at Cochin, preceded by campaign-*jathas* from different parts of the state, a delegates conference presided over by A.K. Gopalan and participated by more than 3,000 delegates, and a mass rally. Following it the CPI(M) organised a mass agitation for 80 days during May-August and 47 days during November-December. The agitation this time was mainly by identifying the surplus and government waste lands and establishing people's right on them.[2]

[1] For details of the convention and the subsequent agitation, see, Gopalan, 1975:114–42; also Oommen, 1985:124–31.

[2] For details of the convention and the follow-up agitation, see, Gopalan, 1975:160–223; also Herring, 1983:203–4; Oommen, 1985:127–36.

True to the realisation of the CPI(M) that mass action was inevitable for speedy and effective implementation of any land reforms, these agitations had significant effects on the KLRAA as will be evident from a discussion of the process and progress of its implementation. But as a backdrop, it is necessary to look at the scope of the Act.

## Scope

It was mainly through three schemes that the KLRAA envisaged achieving its objectives. These were (*a*) conferment of ownership rights of the tenanted lands on the cultivating tenants; (*b*) grant of optional rights to occupants of homesteads (*kudikidappukar*) to purchase their homesteads with appurtenant lands; and (*c*) imposition of ceiling on holdings, acquisition of surplus lands, and their distribution among the landless and the land-poor persons owning or holding less than one acre of land, especially those from the socially and economically backward classes.

For the implementation of the first scheme, invoking section 72 of the Act, the government vested in it, on 1 January 1970, all right, title, and interest of the landowners and intermediaries, including religious, charitable, or educational institutions, on lands held by cultivating tenants except those on which resumption applications were pending under the Act.

For the rights so vested the landowners and intermediaries were entitled to receive as compensation directly from the government the purchase price payable by the cultivating tenants. It was the aggregate of sixteen times the fair rent of the holding plus the value of structures, wells, and embankments of a permanent nature, and one half of the value of timber trees belonging to the landowner and intermediary, not exceeding another sixteen times the fair rent.[3] Religious,

[3] In an interview in May 1979 A.V. Kunhambu, one of the veteran Marxist leaders and former president of the *Kerala Karshaka Sangham*, informed that landlords in Kerala had for a long time led a leisurely and comfortable life by exploiting their tenants, without shedding a drop of sweat or straining a nerve. Therefore, while admitting that the compensation paid was too low, he was of the opinion that his party would have preferred outright confiscation to the payment of even this nominal compensation and that this could not be done for various reasons.

charitable, or educational institutions of a public nature were given the option of getting an annuity in perpetuity of a sum equivalent to the fair rent that they would have obtained from their tenanted lands. Small-holders losing land through the Act for compensation not exceeding Rs. 2,000 were entitled to receive a solatium of up to Rs. 500 from the Agriculturist Rehabilitation Fund constituted under the Act.

While the rights vested in the government were for subsequent assignment to the cultivating tenants, even before such assignment they became *de facto* owners: from the very day of vesting they were not required to pay any rent to either the landowner or intermediary, or even the government, though they were liable to pay the land revenue. For securing the *de jure* ownership through purchase certificates, they had the option to pay the purchase price either in sixteen equal annual instalments or 75 per cent of it in a lump, or settle the matter with the landowners directly on a mutually acceptable basis. The purchase price was treated as debt to the government and as such default in payment entailed only payment of 4.5 per cent annual interest and not forfeiture of ownership rights. Possession alone was made sufficient to claim a valid tenancy, and on the basis of such possession certain categories of cultivators were deemed as tenants eligible to claim ownership rights, so that courts were bound to uphold their rights once they proved possession of land. To help them, the draft record of rights prepared to facilitate the implementation of section 72 of the Act was made a presumptive evidence of possession.

Under the second scheme *kudikidappukar* had the option to purchase from their landowners 3 cents of their homesteads in cities or major municipalities, 5 cents in other municipalities, and 10 cents in panchayats or townships. They were required to pay only 25 per cent of the market value for such purchase, and only half of it if the landowner was in possession of land in excess of the ceiling area. Out of this purchase price one half was subsidised from the *kudikidappukar*'s Benefit Fund constituted under the Act, while the other half was payable in twelve equal annual instalments. To ensure that the landowners did not deprive them of their rights by denying the fact of oral permission granted for occupation, any person who was in possession of any *kudikidappu* from 16 August 1968 to 1 January 1970 was deemed to be a *kudikidappukaran* under the Act.

Under the third scheme the ceiling area fixed was 5 standard acres (or 6 to 7.5 acres) for an adult unmarried person or a family consisting of a sole surviving member, 10 standard acres (12 to 15 acres) for a family of up to five members, with one additional acre for every additional member, subject to an overall ceiling of 15 standard acres (20 acres).[4] It confined exemptions under plantations to tea, coffee, cocoa, rubber, cardamom, or cinnamon, cancelled the earlier exemptions of cashew estates, arecanut, coconut, and pepper gardens, uncultivable waste lands, and *kayal* lands, and confined other exemptions to mainly private forests, and lands belonging to religious, charitable, or educational institutions of a public nature.

## Provisions

With a view to making these schemes—especially the first two—effective, the Act entrenched them with an array of allied provisions some of which are already evident. These provisions extended the benefit of the Act to the maximum number of deserving persons, lessened their legal, procedural, and financial burdens in obtaining the benefits, ensured, by subjecting the process of implementation to public gaze through public notices and such other means that the benefits reached the intended beneficiaries, and plugged some of the serious loopholes in the principal Act.

Thus, related to the first scheme were the provisions for broadening the definition of cultivating tenant, granting fixity of tenure to every tenant, entitling tenants dispossessed on or after 1 April 1964 to restoration of possession, discharging all

[4] To begin with, the compensation payable for surplus lands taken over was 55 per cent of the market value. But by the Kerala Land Reforms (Amendment) Act 25 of 1971, it was substantially reduced. The amendment provided for certain rates of compensation depending upon the nature and class of lands taken over, for scaling down the compensation amount if it exceeded Rs. 1 lakh, and for fixing the maximum compensation payable at Rs. 2 lakh.

rent arrears on payment of the rent for a stipulated period, shifting the onus of proof of tenancy from tenants to landlords, permitting surrender of land by tenants only in favour of the government, prohibiting entry of landowners on lands abandoned by tenants, narrowing substantially the definition of small-holder,[5] restricting severely the land owner's right to resumption, entitling tenants to receive compensation for the lands resumed, and prohibiting retroactively (from 1 April 1964) the creation of new tenancies.

The broadening of the definition of cultivating tenant[6] was mainly by elaborating the category of 'deemed tenants' to accommodate even those 'honestly believing to be tenants,' those occupying land under leases granted by 'incompetent persons' (such as minors), those who had paid amounts for occupation of lands, and those occupying private forests or unsurveyed lands.

The rent payable for discharging all rent arrears was of one year (or the actual arrears whichever was less) by tenants holding up to 5 acres of land, two years by those holding 5 to 10 acres, and 3 years by bigger tenants.

The shifting of the onus of proof of tenancy, that is, instead of cultivating tenants being forced to prove their tenancy status, the landlords being required to prove the opposite, was an extremely important provision to prevent anticipatory evictions and the denial of tenants' rights by landlords, especially because of the widespread absence of record of rights in the state.

While the resumption of *kudiyiruppus* (that is, homestead lands larger and other than *kudikidappus*) and lands from Scheduled Castes and Scheduled Tribes was not allowed at

[5] Under the Act a small-holder is a person who does not have a right in land exceeding 8 standard acres or 10 acres, as owner, intermediary, or cultivating tenant, or as two or more of these, the extent of non-resumable land in whose possession as owner, cultivating tenant, or both does not exceed 2.5 standard acres or 4 acres. See, Government of Kerala, 1977, Vol. 1:23.

[6] Under the Act a cultivating tenant is a tenant who is in actual possession of, and is entitled to cultivate, the land comprised in his holding, solely by his own labour, or with the help of family members, hired labourers, or both, or personally direct or supervise such cultivation. See Government of Kerala, 1977, Vol. 1:2–3, 34–47.

all, resumptions allowed otherwise, under extremely restricted conditions and only within six months from the commencement of the Act, were: (*a*) the whole or part of the holding required for the extension of places of public religious worship (section 14); (*b*) up to 20 cents for construction of residential buildings (section 15); (*c*) the whole or part of the land in excess of ceiling area held by the tenant (section 16); (*d*) not exceeding one-half of the holding for self-cultivation by a small-holder (section 17); and (*e*) agricultural lands interspersed within plantations (section 19).

For the lands resumed, the compensation payable was for improvements made by the tenants, or a solatium of an amount equal to the value of the gross produce from the land resumed for a period of two years, whichever was greater, and both if the lands resumed were from plantations. In the case of lands resumed by small holders for self-cultivation, tenants had the right to choose the portion of the holding allowed to be resumed.

Related to the second scheme were the provisions granting *kudikidappukar* security of tenure, making *kudikidappu* rights heritable, discharging all rent arrears from *kudikidappukar* on payment of one year's rent (or the actual arrears whichever was less), and reducing future rent to not more than Rs. 24 in cities and major municipalities, and Rs. 6 in other areas.[7]

## Machinery

For the implementation of these schemes, the machinery created comprised the Land Board at the apex, Land Tribunals at the base, and Appellate Authorities in the middle.[8] While in the past the Land Board itself supervised

[7] For details of the provisions of the KLRAA, see Government of Kerala, 1977, Vol. 1. For discussion of these provisions, see, Bergmann, 1984:57–59; Herring, 1983:186–89; Nossiter, 1982:293–4; Oommen, 1975:chapters 2 and 3

[8] These were single-member agencies: the Land Board was a member from the Board of Revenue, the Land Tribunal a judicial officer of munsiff's rank or an officer not below the rank of tahsildar, and the Appellate Authority a judicial officer not below the rank of Subordinate Judge or an

and controlled the implementation agencies, because of the enormous increase in work under the 1969 Act the District Collectors, assisted by their Personal Assistants (Land Reforms), were also involved in the implementation work since 1970.

As an appendage of the state's revenue department devoid of any elected representation, the existence of this machinery was in striking contrast to the policy formulations in the KARB. Some of the reasons for this departure seem to have been the direct channel for peasant demands provided by the revenue minister K.R. Gowri as a frontline leader of the KKS, the circulation of the Bill to elicit public opinion, and the actual or apprehended hostility of the central government and the judiciary towards extending statutory powers to popular committees (see Herring, 1983:189–90).

However, as it was only to be expected, there were the mass agitations mentioned earlier and persistent demands from various political parties in the state and the public in general for popular participation in the implementation of the Act, especially in its ceiling provisions. When the implementation commenced in full swing, experience in the working of the Act also revealed the need for such participation. These led to the Kerala Land Reforms (Amendment) Act of 1972 (Act 17 of 1972) published in November 1972 and enforced thence. This amendment provided for the constitution of Village Committees for helping the Land Tribunals in the assignment of ownership rights to tenants, Taluk Land Boards for taking over the functions of the state Land Board in the implementation of the ceiling provisions at the taluk level, and a Land Reforms Review Board to review the progress and lay down guidelines for speedy and effective implementation.

Each Village Committee consisted of the Village Officer as convener, six non-official members, and a chairman elected by the Committee. Each Taluk Land Board consisted of a judicial officer not below the rank of Deputy Collector as chairman, and not more than six non-official members. The

---

officer not below the rank of Deputy Collector. See Government of Kerala, 1977, Vol. 1:173–74.

Land Reforms Review Board consisted of the minister in charge of land reforms as chairman, the Land Board chairman as convener, and six non-official members (see Government of Kerala, 1977, Vol. 1:401–18).

Hard data on the working of the implementation machinery are not available. Nevertheless, data presented in Table 5.1 give a general idea of its composition. When the Act was enforced in 1970 the government constituted as many as 168 Land Tribunals and 4 Appellate Authorities as against the 15 Land Tribunals existing under the principal Act. Of the Tribunals 154 were for implementing the first two schemes; 7 Special Munsiff Land Tribunals were for determining the annuity of public institutions and assigning their tenanted lands; and 7 Munsiff Land Tribunals were for attending to the remaining provisions of the Act concerning resumption of land by landowners, restoration of rights to tenants, fixation of fair rent, cancellation of rent arrears, and so on. Up to 1973 the third scheme of the Act was implemented by the Land Board with the assistance of the District Collectors. In 1973 this work was taken over by the newly formed Taluk Land Boards. Up to 1975 a number of Block Development Officers (BDOs) and Tahsildars (Land Assignment) also functioned as Land Tribunals by holding additional charge. As they could not cope with land reform work in addition to their normal duties, in January 1975 the government replaced them by full-time Tahsildar Land Tribunals. In the same year, it constituted 7 Deputy Collector Land Tribunals to attend to the large number of annuity applications and the related tenancy cases pending with the 7 Special Munsiff Tribunals. During 1975–76 when the implementation of the Act was at its peak, the government further strengthened the machinery by increasing the number of these full-fledged Tribunals.

## Process

In keeping with the complex and enormous nature of the work, the rules framed and procedures followed were both intrinsic and elaborate. This will be evident from a brief description of them.

Table 5.1 *Land Reforms Implementation Machinery by Number of Implementation Agencies, 1970–81*

| *Sl. No.* | *Year* | *Land Board* | *Land Reforms Review Board* | *Taluk Land Board* | *Appellate Authority* | *Land Tribunals* | | | | | | | |
|---|---|---|---|---|---|---|---|---|---|---|---|---|---|
| | | | | | | *Munsiff Land Tribunal* | *Special Munsiff Land Tribunal* | *Deputy Collector Land Tribunal* | *Special Deputy Collector Land Tribunal* | *BDO-cum-Land Tribunal* | *Special Tahsildar (LA)-cum-Land Tribunal* | *Special Tahsildar (LRs) Land Tribunal* | *Total* |
| 1 | 1970 | 1 | — | — | 4 | 7 | 7 | — | — | 85 | 18 | 51 | 168 |
| 2 | 1971 | 1 | — | — | 4 | 7 | 7 | — | – | 83 | 18 | 89 | 204 |
| 3 | 1972 | 1 | — | — | 4 | 7 | 7 | — | — | 89 | 17 | 108 | 228 |
| 4 | 1973 | 1 | 1 | 57 | 4 | 7 | 7 | — | — | 85 | 16 | 112 | 227 |
| 5 | 1974 | 1 | 1 | 57 | 4 | 7 | 7 | 7 | — | — | 2 | 131 | 154 |
| 6 | 1975 | 1 | 1 | 57 | 4 | 6 | 7 | 7 | — | — | 3 | 235 | 258 |
| 7 | 1976 | 1 | 1 | 57 | 4 | 4 | 7 | 7 | 2 | - | 3 | 234 | 257 |
| 8 | 1977 | 1 | 1 | 57 | 4 | 1 | 5 | 9 | 4 | — | — | 197 | 216 |
| 9 | 1978 | 1 | 1 | 57 | 5 | — | 1 | 10 | 5 | — | — | 113 | 129 |
| 10 | 1979 | 1 | 1 | 57 | 5 | — | 1 | 10 | 5 | — | — | 77 | 93 |
| 11 | 1980 | 1 | 1 | 58 | 5 | — | — | 11 | 5 | — | — | 77 | 93 |
| 12 | 1981 | 1 | 1 | 58 | 6 | — | — | 11 | 5 | — | — | 77 | 93 |

Note: Dash indicates nil.
Source: Tabulated from the Administration Reports of the Kerala Land Board and Land Tribunals, 1970–71 to 1981–82.

Under the first scheme, any cultivating tenant entitled to assignment of a tenanted holding applied (in Form A) to the concerned Land Tribunal within the specified time.[9] Any land owner or intermediary of such a holding also applied (in Form C) to the same Tribunal for payment of the compensation due to him after assignment of the holding to the cultivating tenant. In case of mutual consent the land owner, intermediary, and cultivating tenant jointly applied (in Form J) to the Tribunal for an order assigning the holding to the cultivating tenant.[10] Notwithstanding the non-receipt of these applications, the Tribunal initiated *suo motu* proceedings for assigning the holding to the cultivating tenant, and maintained a register (in Form B) of such proceedings.[11] On receipt of an application or after initiating *suo motu* proceedings, the Tribunal published a public notice (in Form D) on the notice board of its office, on that of the concerned village office, and at some prominent part of the holding, and issued a notice (in Form E) individually to each of these persons. Any person interested in the holding but not in receipt of the notice also applied to the Tribunal (in Form F) before the specified date stating his claims or objections.

The public and individual notices of the Tribunal called upon the concerned persons to prefer in writing claims or objections within the specified time, and to appear before the Tribunal on the specified date with all the relevant records to prove their claims or objections. The Tribunal furnished a copy of the public notice along with a statement containing

[9] Initially, the time allowed for applying in Form A was within two years from the commencement of the Act. But the government extended it further right up to 30 June 1976. After this date the Land Tribunals were not authorised to accept applications in Form A, though they were liable to initiate *suo motu* proceedings and accept applications in Forms C and J.

[10] Holdings of cultivating tenants owning land in excess of the ceiling area, or lands owned or leased out by religious, charitable or educational institutions of a public nature claiming annuity instead of compensation were excluded from assignment by mutual consent.

[11] These forms were to contain the village, survey number and extent of the holding, names and addresses of the cultivating tenant, land owner, intermediary, and other persons interested in the holding, particulars of the other lands owned or held by the cultivating tenant or his family, of the tenancy, of the rent paid, and so on. See Government of Kerala, 1977, Vol. 1:110–11; Vol. 2:221–53.

the names and addresses of the persons notified individually, and other related particulars to the Village Committee of the concerned village. The Village Committee advised the Tribunal, after necessary enquiry, on the names and addresses of the persons interested in the holding, and such other matters necessary for the identification of the land comprised in the holding, and its assignment to the cultivating tenant.

After considering the claims and objections, and the advice received from the Village Committee, after such enquiries as it deemed necessary, and after giving a reasonable opportunity of being heard to all the persons notified individually, and to others preferring claims or objections, the Tribunal passed an order assigning the holding. The order specified the extent and survey number of the holding, the purchase price payable by the cultivating tenant, the compensation payable to the landowner, and intermediary, the amount of annuity payable in case of religious, charitable, or educational institutions of a public nature as land owner or intermediary of the holding, and such other particulars. After this order became final the Tribunal issued a certificate of purchase (in Form K) to the cultivating tenant. Thereupon the right, title, and interest of the holding vested in the cultivating tenant became free from all encumbrances created by the land owner or intermediary.

Under the second scheme any *kudikidappukaran* entitled to purchase his *kudikidappu* and appurtenant lands applied (in Form 42) to the Land Tribunal. On receipt of his application the Tribunal issued (in Form 4) notice of such application to the persons interested in the *kudikidappu* land. It also published a copy of this notice on the notice board of its office, on that of the concerned village office, at the office of the concerned local authority, and at the *kudikidappu* land.

After this notice, after necessary enquiries, and after examination of the persons appearing before it, the Tribunal passed such orders on the application as it deemed fit. An order allowing an application specified the extent of the land allowed to be purchased by the *kudikidappukaran*, the purchase price payable, the amount due to the persons interested in the land, and so on. After this order became final, the

Tribunal issued a certificate of purchase (in Form 43). Thereupon the right, title, and interest of the land vested in the *kudikidappukaran* was free from all encumbrances.

Under the third scheme any person owning or holding land in excess of the ceiling area filed a statement (in Form 1) before the Land Board within three months from the commencement of the Act, and certain others filed statements (in Form 1A) within 75 days from the commencement of the Kerala Land Reforms (Amendment) Act of 1972.[12] These statements intimated the location, extent, and such other particulars of all the lands (including those exempted from ceiling) owned or held, and indicated the lands proposed for surrender.

On receipt, the Land Board transferred these statements to the concerned Taluk Land Board. After consideration of the particulars furnished in the statements, the verification reports, and such other relevant matters, the Taluk Land Board prepared a draft statement (in Form 2) of the lands to be surrendered. It published this draft statement on the notice board of its office, and the offices of the concerned District Collector, Taluk Tahsildar and the village office. It also served this statement along with a notice (in Form 3) inviting objections to the persons interested in the lands, and if necessary, also published a public notice (in Form 4) in two daily newspapers with wide circulation in the concerned area.

After considering the objections, and the oral and documentary evidence, and after such further verification as necessary, on the date fixed, by order, the Taluk Land Board determined the extent and identity of the lands for surrender. Thereupon it demanded the surrender of the lands by the issue of a notice (in Form 9) to the persons bound to make the surrender. In case of failure to surrender the Taluk Land Board authorised any officer to take possession or assume ownership of the lands. After taking possession of the lands, the Land Board published a notice (in Form 16) inviting applications before a specified date for assignment on registry

[12] According to this amendment, every family consisting of more than one member owning or holding more than 12 acres of land, every unmarried adult person, and every family consisting of a sole surviving member owning or holding more than 12 acres of land had to file the statement.

of the lands taken over, on the notice board of the office of the Taluk Tahsildar, on that of the office of the local authority, and of the village office, and if necessary in leading Malayalam newspapers. On receipt of the applications and after necessary verification, the Land Board prepared a list of persons eligible for assignment of the lands, showing therein the extent of land (not exceeding one acre) that could be assigned to each applicant. In the assignment of the land, the Land Board gave first preference to the *kudikidappukar* on the land. It assigned the remaining lands to landless agricultural labourers and small-holders, with about 87.5 per cent of such lands available for assignment in a taluk to landless agricultural labourers belonging to Scheduled Castes, Scheduled Tribes, and such other socially and economically backward classes.

Any person aggrieved by any order of the Land Tribunal on the first two schemes appealed within sixty days of such an order to the Appellate Authority. Any person aggrieved by any final order passed in an appeal against any order of the Land Tribunal, the Land Board or Taluk Land Board, preferred a petition to the High Court within 45 days of such order on the ground that the case was decided arbitrarily, or failed to decide any question of law.

Apart from the rules and procedures, the implementation of the three schemes involved other important processes, as revealed by the administration reports of the Kerala Land Board and Land Tribunals. Important among them were the camp sittings held by the Land Tribunals outside their headquarters, whenever these were considered necessary for the convenience of the concerned parties; meetings and conferences of the Land Tribunals and other revenue officers for reviewing the progress and redressing the problems in the working of the Act; public meetings for explaining the salient features of the Act and for eliciting public support and cooperation for its smooth implementation; public functions at different places for issuing certificates of purchase to cultivating tenants and *kudikidappukar* and distribution of surplus lands, focussing attention on the important provisions of the Act; and public distribution of pamphlets printed in simple Malayalam on the various provisions of the Act for educating the public about their rights and obligations.

## Implementation

The implementation of the KLRAA which commenced on 1 January 1970 was virtually over by 31 March 1982. By this date the implementation agencies had disposed of 99.8 per cent of the total applications for assignment of ownership rights to cultivating tenants, 99.1 per cent of the total applications for purchase of *kudikidappus*, and 97.2 per cent of the total land ceiling returns. They had also issued certificates of purchase in 99.2 per cent of the allowed disposals for assignment of ownership rights and 95 per cent of the allowed disposals for purchase of *kudikidappus*. Data presented in Tables 5.2 to 5.6 bring out the main features of implementation.

### *Relative Progress*

In striking contrast to the first and third schemes, implementation of the second scheme was relatively rapid in the initial years. Under this scheme about 76 per cent of the total receipts of applications were during 1970–74, with about 65 per cent of them in the first year itself. About 75 per cent of the total disposals, and 52 per cent of the total purchase certificates issued were also during these five years (Tables 5.4 and 5.5). Under the first scheme only about 30 per cent of the total receipts, 16 per cent of the total disposals, and 8 per cent of the total purchase certificates issued were during 1970–74 (Tables 5.2, 5.3, and 5.5). Under the third scheme only about 14 per cent of the total area ordered for surrender, 19 per cent of the total area taken over, and 5 per cent of the total area distributed were during these five years (Table 5.6).

The relatively rapid implementation of the second scheme in the initial stage itself has to be explained by the land-grab agitation of 1970. The implementation of the first and third schemes continued to be slow right up to 1975. It was only during the national Emergency that the implementation of these schemes made real headway. Under the first scheme, about 57 per cent of the total receipts of applications, 60 per cent of the total disposals, and 57 per cent of the total

Table 5.2 *Receipt of Applications for Assignment of Ownership Rights to Tenants, 1970–81 (Applications in Number and as Percentage of Cumulative Total)*

| Sl. No. | Year | By Tahsildar Tribunals | | | | By Special Tribunals | Grand Total (Cols 6+7) |
|---|---|---|---|---|---|---|---|
| | | Travancore | Cochin | Malabar | Total | | |
| 1 | 2 | 3 | 4 | 5 | 6 | 7 | 8 |
| 1 | 1970 | 6,177 | 35,790 | 93,229 | 1,35,196 | — | 1,35,196 |
| | | 3.1 | 8.2 | 3.8 | 4.4 | | 3.6 |
| 2 | 1971 | 12,828 | 45,286 | 1,73,621 | 2,31,735 | — | 2,31,735 |
| | | 6.5 | 10.4 | 7.1 | 7.5 | | 6.2 |
| 3 | 1972 | 17,875 | 39,473 | 1,40,205 | 1,97,553 | — | 1,97,553 |
| | | 9.0 | 9.0 | 5.8 | 6.4 | | 5.3 |
| 4 | 1973 | 37,285 | 45,437 | 2,14,697 | 2,97,419 | 81,766 | 3,79,185 |
| | | 18.8 | 10.4 | 8.8 | 9.7 | 12.5 | 10.2 |
| 5 | 1974 | 18,906 | 26,508 | 1,10,890 | 1,56,304 | 17,271 | 1,73,575 |
| | | 9.5 | 6.1 | 4,6 | 5.1 | 2.6 | 4.7 |
| 6 | 1975 | 53,916 | 1,16,292 | 10,89,013 | 12,59,221 | 2,56,849 | 15,16,070 |
| | | 27.2 | 26.6 | 44.7 | 41.0 | 39.2 | 40.7 |

**Table 5.2 *Contd.***

| *Sl. No.* | *Year* | *By Tahsildar Tribunals* | | | | *By Special Tribunals* | *Grand Total (Cols 6+7)* |
|---|---|---|---|---|---|---|---|
| | | *Travancore* | *Cochin* | *Malabar* | *Total* | | |
| 1 | 2 | 3 | 4 | 5 | 6 | 7 | 8 |
| 7 | 1976 | 26,002 | 79,407 | 3,59,640 | 4,65,049 | 1,34,409 | 5,99,458 |
| | | 13.1 | 18.2 | 14.8 | 15.1 | 20.5 | 16.1 |
| 8 | 1977–1981 | 25,216 | 48,437 | 2,53,714 | 3,27,367 | 1,64,512 | 4,91,879 |
| | | 12.7 | 11.1 | 10.4 | 10.7 | 25.1 | 13.2 |
| | Total | 1,98,205 | 4,36,630 | 24,35,009 | 30,69,844 | 6,54,807 | 37,24,651 |
| | | 100.0 | 100.0 | 100.0 | 100.0 | 100.0 | 100.0 |
| | | 6.5 | 14.2 | 79.3 | 100.0 | | |

**Note:** Dash indicates nil. Figures for 1970 are from 1 January 1970 to 31 March 1971. Those for the remaining years are for the calendar year beginning on 1 April, and ending on 31 March. Figures for Travancore are the aggregate of the figures for Trivandrum, Quilon, Alleppey, Kottayam and Idikki districts. Those for Cochin and Malabar are the aggregate of the figures for Ernakulam and Trichur districts, and Palghat, Malappuram, Kozhikode, and Cannanore districts respectively. Due to some unexplained discrepancy in the 1977 figures for Cannanore district, data from 1977 to 1981 are not tabulated year-wise. As the receipts and disposals by Special Tribunals (annuity cases) are not available district-wise, these could not be tabulated region-wise.

**Source:** Tabulated from the administration reports of the Kerala Land Board and Land Tribunals, from 1970–71 to 1981–82.

Table 5.3 *Disposal of Applications for Assignment of Ownership Rights to Tenants, 1970–81*

| Sl. No. | Year | Total Disposals | | | | | | Allowed Disposals | | | | | |
|---|---|---|---|---|---|---|---|---|---|---|---|---|---|
| | | By Tahsildar Tribunals | | | | By Special Tribunals | Grand Total (Cols 6+7) | By Tahsildar Tribunals | | | | By Special Tribunals | Grand Total (Cols 12+13) |
| | | Travancore | Cochin | Malabar | Total | | | Travancore | Cochin | Malabar | Total | | |
| 1 | 2 | 3 | 4 | 5 | 6 | 7 | 8 | 9 | 10 | 11 | 12 | 13 | 14 |
| 1 | 1970 | 1,235 | 2,941 | 15,071 | 19,247 | — | 19.247 | 462 | 2,514 | 5,793 | 8,769 | — | 8,769 |
| | | 0.6 | 0.7 | 0.6 | 0.6 | | 0.5 | 37.4 | 85.5 | 38.4 | 45.6 | | 45.6 |
| 2 | 1971 | 5,016 | 24,015 | 60,623 | 89,654 | — | 89,654 | 2,239 | 18,403 | 45,001 | 65,643 | | 65,643 |
| | | 2.5 | 5.5 | 2.5 | 2.9 | | 2.4 | 44.6 | 76.6 | 74.2 | 73.2 | — | 73.2 |
| 3 | 1972 | 9,409 | 29,805 | 79,995 | 1,19,209 | — | 1,19,209 | 4,317 | 21,555 | 65,840 | 91,712 | — | 91,712 |
| | | 4.8 | 6.8 | 3.3 | 3.9 | | 3.2 | 45.9 | 72.3 | 82.3 | 76.9 | | 76.9 |
| 4 | 1973 | 27,668 | 53,254 | 1,25,858 | 2,06,780 | 17,286 | 2,24,066 | 11,055 | 40,073 | 87,870 | 1,38,998 | 12,730 | 1,51,728 |
| | | 14.0 | 12.2 | 5.2 | 6.7 | 2.7 | 6.0 | 40.0 | 75.2 | 69.8 | 67.2 | 73.6 | 67.7 |
| 5 | 1974 | 16,229 | 32,194 | 98,730 | 1,47,153 | 9,614 | 1,56,767 | 6,071 | 19,090 | 65,736 | 90,897 | 7,872 | 98,769 |
| | | 8.2 | 7.4 | 4.1 | 4.8 | 1.5 | 4.2 | 37.4 | 59.3 | 66.6 | 61.8 | 81.9 | 63.0 |
| 6 | 1975 | 52,637 | 67,212 | 5,74,233 | 6,94,082 | 1,29,087 | 8,23,169 | 24,857 | 44,673 | 3,94,990 | 4,64,520 | 1,04,331 | 5,68,851 |
| | | 26.7 | 15.4 | 23.6 | 22.6 | 19.9 | 22.2 | 47.2 | 66.5 | 68.8 | 66.9 | 80.8 | 69.1 |

**Table 5.3 Contd.**

| Sl. No. | Year | Total Disposals | | | | | | Allowed Disposals | | | | | |
|---|---|---|---|---|---|---|---|---|---|---|---|---|---|
| | | By Tahsildar Tribunals | | | | By Special Tribunals | Grand Total (Cols 6+7) | By Tahsildar Tribunals | | | | By Special Tribunals | Grand Total (Cols 12+13) |
| | | Travancore | Cochin | Malabar | Total | | | Travancore | Cochin | Malabar | Total | | |
| 1 | 2 | 3 | 4 | 5 | 6 | 7 | 8 | 9 | 10 | 11 | 12 | 13 | 14 |
| 7 | 1976 | 55,223 | 1,61,331 | 9,66,416 | 11,82,970 | 2,39,152 | 14,22,122 | 26.625 | 99,237 | 6,87,166 | 8,13.028 | 1,84,719 | 9.97.747 |
| | | 28.0 | 37.0 | 39.7 | 38.6 | 36.8 | 38.3 | 48.2 | 61.5 | 71.1 | 68.7 | 77.2 | 70.2 |
| 8 | 1977 | 30,059 | 65,308 | 5,11,260 | 6,06,627 | 2,54,794 | 8,61,421 | 11,383 | 39,511 | 3,16,245 | 3,67,139 | 1,33,057 | 5,00,196 |
| | 1981 | 15.2 | 15.0 | 21.0 | 19.8 | 39.2 | 23.2 | 37.9 | 60.5 | 61.9 | 60.5 | 52.2 | 58.1 |
| | Total | 1,97,476 | 4,36,060 | 24,32,186 | 30,65,722 | 6,49,933 | 37,15,655 | 87,009 | 2,85.056 | 16,68,641 | 20,40,706 | 4,42,709 | 24,83,415 |
| | | 100.0 | 100.0 | 100.0 | 100.0 | 100.0 | 100.0 | 44.1 | 65.4 | 68.6 | 66.6 | 68.1 | 66.8 |
| | | 6.4 | 14.2 | 79.3 | 100.0 | | | 4.2 | 14.0 | 81.8 | 100.0 | | |

**Note:** Total disposals in number and as a percentage of cumulative total. Allowed disposals in number and as a percentage of total yearly disposals. Other details as in Table 5.2.

**Source:** Tabulated from the administration reports of the Kerala Land Board and Land Tribunals, from 1970–71 to 1981–82.

**Table 5.4** *Receipt and Disposal of Kudikidappu Applications, 1970–81*

| Sl. No. | Year | Receipt | | | | Total Disposals | | | | Allowed Disposals | | | |
|---|---|---|---|---|---|---|---|---|---|---|---|---|---|
| | | Travancore | Cochin | Malabar | Total | Travancore | Cochin | Malabar | Total | Travancore | Cochin | Malabar | Total |
| 1 | 1970 | 1,05,491 | 83,776 | 30,730 | 2,19,997 | 41,615 | 33,853 | 11,694 | 87,162 | 29,301 | 29,557 | 6,458 | 65,336 |
| | | 50.4 | 52.8 | 39.2 | 49.3 | 20.1 | 21.5 | 15.0 | 19.7 | 70.4 | 87.4 | 55.2 | 75.0 |
| 2 | 1971 | 21,392 | 22,684 | 8,112 | 52,188 | 57,442 | 40,498 | 11,866 | 1,09,806 | 34,857 | 23,030 | 6,646 | 64,533 |
| | | 10.2 | 14.3 | 10.4 | 11.7 | 27.7 | 25.8 | 15.2 | 24.8 | 60.7 | 56.9 | 56.0 | 58.8 |
| 3 | 1972 | 15,009 | 14,274 | 11,006 | 40,289 | 36,933 | 42,403 | 20,044 | 99,380 | 22,690 | 22,757 | 11,879 | 57,326 |
| | | 7.2 | 9.0 | 14.1 | 9.0 | 17.8 | 27.0 | 25.7 | 22.5 | 61.4 | 53.7 | 59.3 | 57.7 |
| 4 | 1973 | 7.607 | 5,254 | 3,857 | 16,718 | 9,806 | 5,842 | 7,806 | 23,454 | 5,711 | 4,271 | 5,007 | 14,989 |
| | | 3.6 | 3.3 | 4.9 | 3.7 | 4.7 | 3.7 | 10.0 | 5.3 | 58.2 | 73.1 | 64.2 | 63.9 |
| 5 | 1974 | 4,702 | 4,045 | 2,461 | 11,208 | 4,353 | 3,503 | 2,869 | 10,725 | 2,560 | 2,261 | 1,814 | 6,635 |
| | | 2.2 | 2.6 | 3.1 | 2.5 | 2.1 | 2.2 | 3.7 | 2.4 | 58.8 | 64.5 | 63.2 | 61.9 |
| 6 | 1975 | 20,765 | 9,463 | 8,835 | 39,063 | 20,184 | 9,036 | 7,467 | 36,687 | 12,653 | 5,095 | 3,846 | 21,594 |
| | | 9.9 | 6.0 | 11.3 | 8.8 | 9.8 | 5.7 | 9.6 | 8.3 | 62.7 | 56.4 | 51.5 | 58.9 |
| 7 | 1976 | 12,225 | 8,791 | 5,108 | 26,124 | 16,448 | 12,094 | 7,274 | 35,816 | 10,479 | 5,626 | 3,909 | 20,014 |
| | | 5.8 | 5.5 | 6.5 | 5.9 | 7.9 | 7.7 | 9.3 | 8.1 | 63.7 | 46.5 | 53.7 | 55.9 |
| 8 | 1977 | 4,379 | 2,811 | 2,664 | 9,854 | 4,455 | 3,253 | 2,966 | 10,674 | 2,366 | 1,752 | 1,639 | 5,757 |
| | | 2.1 | 1.8 | 3.4 | 2.2 | 2.2 | 2.1 | 3.8 | 2.4 | 53.1 | 53.9 | 55.3 | 53.9 |
| 9 | 1978 | 7,362 | 2,645 | 1,892 | 11,899 | 5,913 | 2,780 | 2,176 | 10,869 | 3,446 | 1,679 | | |
| | | 3.5 | 1.7 | 2.4 | 2.7 | 2.9 | 1.8 | 2.8 | 2.5 | 58.3 | 60.4 | | |

Table 5.4 *Contd.*

| *Sl. No.* | *Year* | *Receipt* | | | | *Total Disposals* | | | | *Allowed Disposals* | | | |
|---|---|---|---|---|---|---|---|---|---|---|---|---|---|
| | | *Travancore* | *Cochin* | *Malabar* | *Total* | *Travancore* | *Cochin* | *Malabar* | *Total* | *Travancore* | *Cochin* | *Malabar* | *Total* |
| 10 | 1979 | 3,706 | 1,755 | 1,251 | 6,712 | 3,603 | 1,344 | 1,166 | 6,113 | 2,017 | 913 | 2,058 | 10,113 |
| | | 1.8 | 1.1 | 1.6 | 1.5 | 1.7 | 0.9 | 1.5 | 1.4 | 56.0 | 67.9 | 61.6 | 59.6 |
| 11 | 1980–81 | 6,514 | 3,032 | 2,379 | 11,925 | 6,291 | 2,641 | 2,516 | 11,448 | 3,341 | 1,899 | 1,561 | 6,821 |
| | | 3.1 | 1.9 | 3.0 | 2.7 | 3.0 | 1.7 | 3.2 | 2.6 | 53.1 | 71.9 | 62.0 | 59.6 |
| | Total | 2,09,152 | 1,58,530 | 78,295 | 4,45,977 | 2,07,043 | 1,57,247 | 77,844 | 4,42,134 | 1,29,421 | 98,860 | 44,817 | 2,73,118 |
| | | 100.0 | 100.0 | 100.0 | 100.0 | 100.0 | 100.0 | 100.0 | 100.0 | 62.5 | 62.9 | 57.6 | 61.8 |
| | | 46.9 | 35.5 | 17.6 | 100.0 | 46.8 | 35.6 | 17.6 | 100.0 | 47.4 | 36.2 | 16.4 | 100.0 |

**Note:** Due to some unexplained discrepancy in the 1978 cumulative figures for Cannanore district, the 1978–79 data for Malabar are not tabulated separately. Other details are as in Table 5.2

**Source:** Tabulated from the administration reports of the Kerala Land Board and Land Tribunals from 1971–72 to 1981–82.

**Table 5.5** ***Purchase Certificates Issued to Cultivating Tenants and Kudikidappukar, 1970–81 (Certificates in Number and Percentage)***

| *Sl. No.* | *Year* | *Certificates Issued to* | |
|---|---|---|---|
| | | *Tenants* | *Kudikidappukar* |
| 1 | Up to 1972 | 81,825 | 80,329 |
| | | 3.3 | 31.0 |
| 2 | 1973 | 53,207 | 45,773 |
| | | 2.2 | 17.6 |
| 3 | 1974 | 49,132 | 9,815 |
| | | 2.0 | 3.8 |
| 4 | 1975 | 3,34,259 | 34,216 |
| | | 13.6 | 13.2 |
| 5 | 1976 | 10,75,976 | 45,135 |
| | | 43.7 | 17.4 |
| 6 | 1977 | 5,66,331 | 21,669 |
| | | 23.0 | 8.4 |
| 7 | 1978 | 1,91,250 | 10,047 |
| | | 7.8 | 3.9 |
| 8 | 1979 | 50,698 | 5,387 |
| | | 2.1 | 2.1 |
| 9 | 1980–1981 | 61,598 | 7,039 |
| | | 2.5 | 2.7 |
| | Total | 24,64,276 | 2,59,410 |
| | | 100.0 | 100.0 |

**Source:** Tabulated from the administration reports of Kerala Land Board and Land Tribunals, 1970–71 to 1981–82.

Table 5.6 *Acquisition and Distribution of Surplus Lands, 1970-81 (Area in Acres)*

| Sl. No. | Year | Area Ordered for Surrender | | | | Area Taken Over | | | | Area Distributed | | | |
|---|---|---|---|---|---|---|---|---|---|---|---|---|---|
| | | Travancore | Cochin | Malabar | Total | Travancore | Cochin | Malabar | Total | Travanco. e | Cochin | Malabar | Total |
| 1 | 1971 | 251 | 7 | 699 | 957 | — | — | 316 | 316 | — | — | 230 | 230 |
| | | 0.6 | 0.1 | 0.6 | 0.6 | | | 0.5 | 0.4 | | | 0.6 | 0.4 |
| 2 | 1972 | 1,873 | 270 | 7,307 | 9,450 | 1 185 | 259 | 5,067 | 6,511 | 91 | — | 872 | 963 |
| | | 4.6 | 2.1 | 6.7 | 5.8 | 6.2 | 4.7 | 8.4 | 7.7 | 0.8 | | 2.4 | 1.8 |
| 3 | 1973 | 3,491 | 56 | 1,849 | 5,396 | 3,352 | 3 | 2,485 | 5,840 | 91 | 39 | 611 | 741 |
| | | 8.7 | 0.4 | 1.7 | 3.3 | 17.5 | 0.1 | 4.1 | 6.9 | 0.8 | 1.0 | 1.6 | 1.4 |
| 4 | 1974 | 1,703 | 503 | 5,514 | 7,720 | 457 | 181 | 2,545 | 3,183 | 180 | — | 349 | 529 |
| | | 4.2 | 3.9 | 5.0 | 4.7 | 2.4 | 3.3 | 4.2 | 3.8 | 1.5 | | 0.9 | 1.0 |
| 5 | 1975 | 4,613 | 3,690 | 23,191 | 31,494 | 1,843 | 626 | 8,729 | 11,198 | 2,663 | 664 | 5,576 | 8,903 |
| | | 11.4 | 28.4 | 21.2 | 19.3 | 9.6 | 11.4 | 14.6 | 13.2 | 22.3 | 17.2 | 15.0 | 16.8 |
| 6 | 1976 | 16,259 | 4,804 | 36,674 | 57,737 | 6,380 | 2,483 | 21,233 | 30,096 | 5,395 | 1,477 | 13,018 | 19,890 |
| | | 40.3 | 37.0 | 33.5 | 35.5 | 33.4 | 45.2 | 35.4 | 35.6 | 45.3 | 38.2 | 35.1 | 37.6 |
| 7 | 1977 | 6,515 | 1,620 | 12,551 | 20,686 | 2,342 | 923 | 6,702 | 9,967 | 1,576 | 831 | 8,267 | 10,674 |
| | | 16.2 | 12.5 | 11.5 | 12.7 | 12.3 | 16.8 | 11.2 | 11.8 | 13.2 | 21.5 | 22.3 | 20.2 |
| 8 | 1978 | 1,998 | 1.163 | 6,173 | 9,334 | 941 | 375 | 4,051 | 5.367 | 585 | 220 | 3,089 | 3,894 |
| | | 5.0 | 9.0 | 5.6 | 5.7 | 4.9 | 6.8 | 6.8 | 6.3 | 4.9 | 5.7 | 8.3 | 7.4 |
| 9 | 1979 | 1,430 | 193 | 3,589 | 5,212 | 856 | 363 | 2,782 | 4,001 | 265 | 449 | 1,780 | 2,494 |
| | | 3.6 | 1.5 | 3.3 | 3.2 | 4.5 | 6.6 | 4.6 | 4.7 | 2.2 | 11.6 | 4.8 | 4.7 |

Table 5.6 *Contd.*

| Sl. No. | Year | Area Ordered for Surrender | | | | Area Taken Over | | | | Area Distributed | | | |
|---|---|---|---|---|---|---|---|---|---|---|---|---|---|
| | | *Travancore* | *Cochin* | *Malabar* | *Total* | *Travancore* | *Cochin* | *Malabar* | *Total* | *Travancore* | *Cochin* | *Malabar* | *Total* |
| 10 | 1980 | 1,105 | 270 | 5,487 | 6,862 | 1,303 | 145 | 3,032 | 4,480 | 675 | 84 | 2,289 | 3,048 |
| | | 2.7 | 2.1 | 5.0 | 4.2 | 6.8 | 2.6 | 5.1 | 5.3 | 5.7 | 2.2 | 6.2 | 5.8 |
| 11 | 1981 | 1,106 | 398 | 6,428 | 7,932 | 451 | 140 | 3,044 | 3,635 | 400 | 103 | 990 | 1,493 |
| | | 2.7 | 3.1 | 5.9 | 4.9 | 2.4 | 2.5 | 5.1 | 4.3 | 3.3 | 2.7 | 2.7 | 2.8 |
| | Total | 40,344 | 12,974 | 1,09,462 | 1,62,780 | 19,110 | 5,498 | 59,986 | 84,594 | 11,921 | 3,867 | 37,071 | 52,859 |
| | | 100.0 | 100.0 | 100.0 | 100.0 | 100.0 | 100.0 | 100.0 | 100.0 | 100.0 | 100.0 | 100.0 | 100.0 |
| | | 24.8 | 8.0 | 67.2 | 100.0 | 22.6 | 6.5 | 70.9 | 100.0 | 22.6 | 7.3 | 70.1 | 100.0 |

**Note:** Dash indicates nil.
**Source:** Tabulated from the administration reports of the Kerala Land Board and Land Tribunals, from 1970–71 to 1981–82.

purchase certificates issued were during 1975–76 (Tables 5.2, 5.3, and 5.5). Similarly, under the third scheme about 55 per cent of the area ordered for surrender, 49 per cent of the total area taken over, and 54 per cent of the total area distributed were during these two years. In fact, the total disposals during the first 13 months of the Emergency were almost double those in the preceding 68 months. The monthly disposals by the Land Tribunals also shot up from about 16,000 cases to about 1.2 lakh cases, and the average disposals per Tribunal from 109 to 489.[13] By the time the Emergency was lifted the implementation of the Act was also virtually over, though keeping in view the Emergency excesses it is only to be expected that this accelerated pace would have also had a negative bearing on the quality of implementation.[14]

In this connection it will be interesting to know why, after a delay in implementation for nearly half a decade, the CPI-led coalition suddenly activised the implementation and made it time-bound,[15] more so, when it was scared of its partner, the Congress Party, which had all along protected the landed interests and opposed the earlier attempts at land reforms. Perhaps the answer to this has to be found in the political reasons which led to the declaration of the Emergency and to

[13] See *Dharani*, Special Number on Kerala Land Reforms, November 1976:15.

[14] Though hard data are not available, some indications of this negative effect could be gathered during fieldwork in north Malabar from the stories which some *janmis* and tenants had to tell about the callous, careless, and clear-the-deck-type work of the Tribunals. Their complaints were that they were tired of being driven from pillar to post by the Tribunals and that some of them did not get their lands, or their legitimate compensation.

[15] It is interesting to note that during the Emergency, with a view to motivating the land reforms staff at the implementation level, the state Land Board brought out an attractive journal, *Dharani*, from December 1975 to May 1977. Its editor, Malayatoor Ramakrishnan, a doyen of Malayalam literature and chairman of the state Land Board elevated the work of the land reforms staff to an 'Olympic Game' by awarding 'gold' and 'silver' medals to the 'best athletes,' and announcing their scores every month through its columns. Good work evidently meant quick disposal of more and more cases by each Tribunal. And inspired by the 'laurels' of this 'Land Reforms Olympics' many of the Tribunals disposed of more than 2,000 cases a month. It was reported that Cannanore district ranked first in this game, that one of its Tribunals disposed of as many as 2,267 cases in one month, and that this was the state record. See *Dharani*, November 1976:59.

the inclusion of land reforms in the 20-point programme of the prime minister, forcing the unwilling Congress to force such reforms on the equally unwilling state governments.

### *Relative Performance*

Not withstanding the delays and the likely lapses, especially during the Emergency, the implementation of the first two schemes seems to have been fairly efficient and successful, whereas the same cannot be said about the third scheme. Data available up to 31 March 1982 give a general idea of the relative performance of these schemes.

Under the first scheme the total receipts and disposals by the Land Tribunals were 37,24,651 and 37,15,655 (99.8 per cent) applications respectively. Of the disposals 24,83,415 (66.8 per cent) were allowed and the rest were rejected, including cases which were struck off due to errors, duplications, and so on.

About 82 per cent of the total receipts and disposals, and 80 per cent of the total allowed disposals were by the Tahsildar Tribunals, and only the rest were by the Special Tribunals. The district-wise details of the disposals by the former indicate that about 79 per cent of the total receipts and disposals and 82 per cent of the total allowed disposals were in the Malabar region alone. Only about 6 per cent of the total receipts and disposals and 4 per cent of the allowed disposals were in Travancore, and only 14 per cent of all the three were in Cochin (Tables 5.2 and 5.3). These disposals have to be seen against the estimate by the 1966–67 Land Reforms Survey (LRS)—the only statistically significant survey of tenancy in the state on the eve of the implementation of the KLRAA—of the tenancy households in the state about 64 per cent were in Malabar and only about 12 per cent were in Travancore and 23 per cent in Cochin (see Government of Kerala, 1968:133).

The area involved in the allowed disposals up to the end of February 1981, that is, in 24,59,491 or 99 per cent of the total allowed disposals, was estimated by the Land Board as 19,67,593 acres.[16] This closely corresponds to the total leased

[16] As intimated by Kerala Land Board, Trivandrum, letter no. L/Dis/20058/ 81/LB dated 5 May 1981.

in area of 19.2 lakh acres as estimated by the LRS. The LRS had also estimated the total tenant households in the state as 12.7 lakh, so that the average leased-in area per tenant household was 1.5 acres (see Government of Kerala, 1968: 57–59, 133, 273). Assuming that the LRS estimates of 1966 did not vary widely from the actual situation before the commencement of the KLRAA in 1970, and assuming further that the slippage in implementation was not very high, the total allowed disposals and area disposed of by the Tribunals seen in conjunction with the LRS estimates indicate the following: (*a*) the average area per disposal or holding was 0.8 acre; (*b*) since the average area per tenant household according to the LRS was 1.5 acres the average number of allowed disposals or leased-in holding per tenant household was 1.9 (each disposal was for a separate holding); and (*c*) the tenants benefited by the total allowed disposals were about 12.9 lakhs (cf. Herring, 1980:60–61; 1983:211).

Under the second scheme the total receipts and disposals by the Tribunals were 4,45,977 and 4,42,134 (99 per cent) applications respectively. Of the total disposals 2,73,118 (62 per cent) were allowed and the rest were rejected.

In contrast to the first scheme, about 47 per cent of the total receipts, disposals, and allowed disposals were in Travancore, and about 36 per cent of all the three were in Cochin, whereas only about 18 per cent of the total receipts and disposals and 16 per cent of the allowed disposals were in Malabar (Table 5.4). These disposals also have to be seen against the LRS estimate—of the total *kudikidappus* in the state 50 per cent were in Travancore, 27 per cent in Cochin and only 23 per cent in Malabar (see Government of Kerala, 1968:133).

The area involved in the allowed disposals up to the end of February 1981, that is, in 2,69,021 or 98 per cent of the allowed disposals, was estimated by the Land Board as 21,522 acres, that is, an average of 0.08 acre (8 cents) per allowed disposal. The LRS had estimated the number of *kudikidappus* in the state as 3.4 lakhs (see Government of Kerala, 1968:61, 118–19, 127). While the disposals allowed by the Tribunals account for only 80 per cent of this estimate, the remaining 20 per cent should not be construed as slippage. A large number of *kudikidappukars* had fenced in their lands during the land-grab agitation of 1970. This agitation was acclaimed as almost a complete success by its leader A.K. Gopalan himself (see Gopalan, 1975:143). Many of these *kudikidappukar* had refused to submit the required applications to the Tribunals (see Oommen, 1985:126).

Under the third scheme, of the 1,62,780 acres ordered for surrender, 42,048 acres (26 per cent) were involved in court stays, disputes, vested forest lands, and so on, so that only 1,20,732 (74 per cent) acres were available for take over. Of the 84,595 acres taken over 3,909 acres were returned to the owners because of court rulings and other reasons. Of the 80,686 acres (50 per cent) available for distribution 23,110 acres (14 per cent) were distributed among 39,366 persons from Scheduled Castes and Scheduled Tribes, and 26,743 acres (16 per cent) among 45,798 'others,' at the average rate of about 59 cents per person, and 3,006 acres (1.8 per cent) among 3 institutions at the rate of 1,002 acres per institution.[17] The area distributed was only about 32 per cent of that ordered for surrender, and only 46 per cent of even the LRS estimate of 1.15 lakh acres. The latter itself was a gross underestimate of the available surplus because of the provisions in the 1964 Act for higher ceiling and numerous exemptions. As many of the potential surplus holders had either sold or rearranged their holdings by the time of the LRS itself, neither the area ordered for surrender nor the LRS estimate indicates the real surplus.

The oft-quoted estimate of the first communist ministry of 17.5 lakh acres, even if on the high side, or the subsequent estimate of 4 lakh acres after allowing for slippage for about a decade[18] could probably be considered as nearer the mark. Viewed against these estimates the performance of the third scheme has been dismal.

### *Failure of Ceiling*

This poor performance has been the result of a combination

[17] See Administration Report of the Kerala Land Board and Land Tribunals, 1981–82.

[18] See Report of the *Kerala Karshaka Sangham*, presented to the 15th state conference held at Thodupuzha on 9–11 March 1979.

of factors, and calls for some explanation. When the Communist Party came to power in 1957, the landlords in Kerala rightly apprehended a serious threat to their feudal interests in land and made hectic land transfers even as the KARB was on the anvil. The enactment of the 1961 and 1964 Acts, even as its diluted versions hastened the process of further transfers. The LRS data clearly show that most of these transfers were bogus, and made with a view to avoiding the surrender of lands. Of the 3,12,678 ownership transfers reported by the LRS for the ten years 1957–66 and of the 2,15,813 acres transferred, about two-thirds were in the course of just four years. The peak years were 1960, 1962, 1963 and 1964 in the state as a whole and in the Travancore region. They were 1960, 1962 and 1964 in Cochin, and 1957 and 1964 in Malabar. Of all the transfers 70 per cent were through outright sales, 20 per cent through partitions, and 10 per cent through gifts. The area transferred accounted for 68 per cent, 28 per cent and 4 per cent respectively. Transfers through partitions and gifts were mostly confined to the Travancore region. In fact about 94 per cent of all the partition-transfers and 76 per cent of all the gift-transfers in the state were in this region. More importantly, about 74 per cent of all the transfers through all the three modes and 58 per cent of all the area transferred were in it, as against only 8 per cent of the transfers and 28 per cent of the area transferred in Malabar (Tables 5.7 and 5.8). Considering these transfers it is not surprising that of the total surplus lands taken over and distributed in the state the contribution from Travancore was not even one-fourth whereas that from Malabar as almost three-fourths.

Bogus transfers were, however, only one of the important reasons for the disappearance of surplus lands. There were large-scale conversions of lands into plantations. These, and the exemption of plantations from ceiling in the successive land reforms legislations in the state, considerably reduced the scope of land redistribution and social change through land reforms. What is more, there was hardly any implementation of land reforms, especially the ceiling provisions, right up to 1970 due to, first, the indifference of the Congress-coalition, and then the political instability and President's

**Table 5.7 *Ownership Transfers by Number, Type and Region, 1957–66***

| Sl. No. | Year | Travancore | | | | Cochin | | | | Malabar | | | | Grand Total |
|---|---|---|---|---|---|---|---|---|---|---|---|---|---|---|
| | | Sale | Partition | Gift | Total | Sale | Partition | Gift | Total | Sale | Partition | Gift | Total | |
| 1 | 1957 | 9,216 | 2,684 | — | 11,900 | — | — | — | — | 2,530 | — | — | 2,530 | 14,430 |
| | | 6.3 | 4.4 | | 5.2 | | | | | 11.7 | | | 10.3 | 4.6 |
| 2 | 1958 | 8,630 | 2,639 | — | 11,269 | 3,769 | — | 1,381 | 5,150 | 1,501 | 1,228 | — | 2,729 | 19,148 |
| | | 5.9 | 4.4 | | 4.9 | 7.5 | | 23.9 | 8.9 | 6.9 | 100.0 | | 11.1 | 6.1 |
| 3 | 1959 | 2,607 | — | — | 2,607 | 4,807 | — | — | 4,807 | 919 | — | 635 | 1,554 | 8.968 |
| | | 1.8 | | | 1.1 | 9.6 | | | 8.3 | 4.3 | | 35.2 | 6.3 | 2.9 |
| 4 | 1960 | 18.955 | 21.806 | — | 40,761 | 3,745 | 166 | — | 3,911 | 1,805 | — | — | 1,805 | 46,477 |
| | | 13.0 | 36.2 | | 17.7 | 7.5 | 7.0 | | 6.7 | 8.4 | | | 7.3 | 14.9 |
| 5 | 1961 | 10,156 | 212 | — | 10,368 | 3,597 | 166 | — | 3,763 | 1,754 | — | 486 | 2,240 | 16,371 |
| | | 7.0 | 0.3 | | 4.5 | 7.2 | 7.0 | | 6.5 | 8.1 | | 26.9 | 9.1 | 5.2 |
| 6 | 1962 | 20,825 | 805 | — | 21,630 | 6,975 | — | 4,401 | 11,376 | 3,340 | — | — | 3,340 | 36,346 |
| | | 14.2 | 1.3 | | 9.4 | 14.0 | | 76.1 | 19.6 | 15.4 | | | 13.5 | 11.6 |
| 7 | 1963 | 29.147 | 5,536 | 7,064 | 41.747 | 4,937 | 850 | — | 5,787 | 3,840 | — | — | 3,840 | 51,374 |
| | | 19.9 | 9.2 | 29.9 | 18.2 | 9.9 | 35.9 | | 10.0 | 17.8 | | | 15.6 | 16.4 |
| 8 | 1964 | 15,482 | 22,808 | 13,548 | 51,838 | 10,109 | 687 | — | 10,796 | 1,216 | — | 683 | 1,899 | 64,533 |
| | | 10.6 | 37.9 | 57.4 | 22.5 | 20.3 | 29.0 | | 18.6 | 5.6 | | 37.9 | 7.7 | 20.6 |

Table 5.7 *Contd.*

| Sl. No. | Year | Travancore | | | | Cochin | | | | Malabar | | | | Grand Total |
|---|---|---|---|---|---|---|---|---|---|---|---|---|---|---|
| | | *Sale* | *Partition* | *Gift* | *Total* | *Sale* | *Partition* | *Gift* | *Total* | *Sale* | *Partition* | *Gift* | *Total* | |
| 9 | 1965 | 13,776 | 3,661 | 2,684 | 20,121 | 9,277 | — | — | 9,277 | 3,706 | — | — | 3,706 | 33,104 |
| | | 9.4 | 6.1 | 11.4 | 8.8 | 18.6 | | | 16.0 | 17.1 | | | 15.0 | 10.6 |
| 10 | 1966 | 17,327 | 106 | 291 | 17,724 | 2,688 | 498 | — | 3,186 | 1,017 | — | — | 1,017 | 21,927 |
| | | 11.9 | 0.2 | 1.2 | 7.7 | 5.4 | 21.0 | | 5.5 | 4.7 | | | 4.1 | 7.0 |
| | Total | 1,46,121 | 60,257 | 23,587 | 2,29,965 | 49,904 | 2,367 | 5,782 | 58,053 | 21,628 | 1,228 | 1,804 | 24,660 | 3,12,678 |
| | | 100.0 | 100.0 | 100.0 | 100.0 | 100.0 | 100.0 | 100.0 | 100.0 | 100.0 | 100.0 | 100.0 | 100.0 | 100.0 |
| | | 63.5 | 26.2 | 10.3 | 100.0 | 86.0 | 4.1 | 9.9 | 100.0 | 87.0 | 5.0 | 7.3 | 100.0 | 100.0 |
| % of State Total | | 67.1 | 94.4 | 75.7 | 73.5 | 22.9 | 3.7 | 18.5 | 18.6 | 9.9 | 1.9 | 5.8 | 7.9 | 100.0 |

**Note:** Dash indicates nil.

**Source:** Tabulated from Government of Kerala, 1968:285–301.

Table 5.8 *Ownership Transfers by Area, Type and Region, 1957-66 (Area in Acres)*

| *Sl. No.* | *Year* | *Travancore* | | | | *Cochin* | | | | *Malabar* | | | | *Grand Total* |
|---|---|---|---|---|---|---|---|---|---|---|---|---|---|---|
| | | *Sale* | *Partition* | *Gift* | *Total* | *Sale* | *Partition* | *Gift* | *Total* | *Sale* | *Partition* | *Gift* | *Total* | |
| 1 | 1957 | 1,776 | 3,463 | — | 5,239 | — | — | — | — | 8,507 | — | — | 8,507 | 13,746 |
| | | 3.0 | 5.9 | | 4.2 | | | | | 14.5 | | | 14.1 | 6.4 |
| 2 | 1958 | 7,135 | 542 | — | 7,677 | 2,250 | — | 728 | 2,978 | 1,277 | 442 | — | 1,719 | 12,374 |
| | | 12.0 | 0.9 | | 6.2 | 7.9 | | 33.6 | 9.5 | 2.2 | 100.0 | | 2.9 | 5.4 |
| 3 | 1959 | 5,266 | — | — | 5,266 | 2,645 | — | — | 2,645 | 1,138 | — | 70 | 1,208 | 9,119 |
| | | 8.8 | | | 4.2 | 9.3 | | | 8.4 | 1.9 | | 5.8 | 2.0 | 4.2 |
| 4 | 1960 | 9,619 | 31,038 | — | 40,657 | 4,415 | 116 | — | 4,531 | 5,741 | — | — | 5,741 | 50,929 |
| | | 16.2 | 52.7 | | 32.8 | 15.6 | 12.2 | | 14.4 | 9.8 | | | 9.5 | 23.6 |
| 5 | 1961 | 1,481 | 426 | — | 1,907 | 1,177 | 312 | — | 1,489 | 3,700 | — | 1,093 | 4,793 | 8,189 |
| | | 2.5 | 0.7 | | 1.5 | 4.1 | 32.8 | | 4.7 | 6.3 | | 91.2 | 8.0 | 3.8 |
| 6 | 1962 | 7,002 | 3,146 | — | 10,148 | 5,151 | — | 1,438 | 6,589 | 2,480 | — | — | 2,480 | 19,217 |
| | | 11.8 | 5.3 | | 8.2 | 18.2 | | 66.4 | 20.9 | 4.2 | | | 4.1 | 8.9 |
| 7 | 1963 | 10,072 | 2,831 | 1,590 | 14,493 | 1,905 | 279 | — | 2,184 | 26,707 | — | — | 26,707 | 43,384 |
| | | 16.9 | 4.8 | 28.0 | 11.7 | 6.7 | 29.4 | | 6.9 | 45.6 | | | 44.3 | 20.1 |
| 8 | 1964 | 2,777 | 16,309 | 2,283 | 21,369 | 5,614 | 103 | — | 5,717 | 2,065 | — | 36 | 2,101 | 29,187 |
| | | 4.7 | 27.7 | 40.1 | 17.2 | 19.8 | 10.8 | | 18.2 | 3.5 | | 3.0 | 3.5 | 13.5 |

Table 5.8 *Contd.*

| *Sl. No.* | *Year* | Travancore *Sale* | *Partition* | *Gift* | *Total* | Cochin *Sale* | *Partition* | *Gift* | *Total* | Malabar *Sale* | *Partition* | *Gift* | *Total* | *Grand Total* |
|---|---|---|---|---|---|---|---|---|---|---|---|---|---|---|
| 9 | 1965 | 6,215 | 1,036 | 1,771 | 9,022 | 3,136 | — | — | 3,136 | 6,052 | — | — | 6,052 | 18,210 |
| | | 10.4 | 1.8 | 31.1 | 7.3 | 11.1 | | | 10.0 | 10.3 | | | 10.0 | 8.4 |
| 10 | 1966 | 8,149 | 134 | 44 | 8,327 | 2,058 | 140 | — | 2,198 | 933 | — | — | 932 | 11,458 |
| | | 13.7 | 0.2 | 0.8 | 6.7 | 7.3 | 14.7 | | 7.0 | 1.6 | | | 1.5 | 5.3 |
| | Total | 59,492 | 58,925 | 5,688 | 1,24,105 | 28,351 | 950 | 2,166 | 31,737 | 58,599 | 442 | 1,199 | 60,241 | 2,15,813 |
| | | 100.0 | 100.0 | 100.0 | 100.0 | 100.0 | 100.0 | 100.0 | 100.0 | 100.0 | 100.0 | 100.0 | 100.0 | 100.0 |
| | | 47.9 | 47.5 | 4.6 | 100.0 | 90.1 | 3.0 | 6.9 | 100.0 | 97.3 | 0.7 | 2.0 | 100.0 | 100.0 |
| % of State Total | | 40.6 | 97.7 | 62.8 | 57.5 | 19.4 | 1.6 | 23.9 | 14.6 | 40.0 | 0.7 | 13.2 | 27.9 | 100.0 |

**Note:** Dash indicates nil.
**Source:** Tabulated from Government of Kerala, 1968:285–301.

rule in the state. This gave ample opportunities to the land owners to circumvent any future ceiling on their holdings.

When the implementation of land reforms eventually commenced in 1970, the CPI-led coalition, faced with an Act drafted and legislated by its political rival, the CPI(M), did not make any serious effort to implement it. This was especially so in the case of the ceiling provisions, as the CPI was scared of losing the support of the Muslim League and the Kerala Congress, the two mainstay groups of the coalition. This reluctance was despite chief minister Achuta Menon's earlier assertion that had the CPI(M)-led coalition heeded his party's demand and brought into force the ceiling provisions of the 1964 Act immediately after assuming office, all transfers of land in excess of the ceiling after 1963 would have become invalid (Menon, 1969:15–16). And, when implementation of the ceiling provisions began to gain a measure of stability it was soon vitiated by the politics of voluntary transfers.

In fact, both the 1964 Act and the 1969 Act contained ample provisions for voluntary transfers. True, section 84 of the former invalidated all voluntary transfers effected after 1 January 1964, by a family or any of its members, or by an adult unmarried person owning or holding land in excess of the ceiling area. But this section was self-defeating because of its wide scope for exemptions from such invalidation of transfers (*a*) by way of partition; (*b*) on account of natural love and affection; (*c*) in favour of a person who was a tenant of the holding before 27 July 1960, and continued to be so till the date of transfer; or (*d*) in favour of a religious, charitable, or educational institution of a public nature solely for the purpose of the institution. The 1969 Act also retained this section, though from among the above exemptions it deleted the second. What is more, under it the date invalidating voluntary transfers was only 1 July 1969 and not 14 January 1964, so that all transfers made during the intervening six years were valid.

Through the Kerala Land Reforms (Amendment) Act 17 of 1972, introduced after the excess land agitation of 1972, the government deleted the second and fourth exemptions of the 1964 Act with effect from 16 August 1968 itself. But, it also introduced a new clause validating from the same date voluntary transfers by way of gift, by any person owning or holding land in excess of the ceiling area, in favour of his son or daughter or the son or daughter of his pre-deceased son or daughter. It also interpreted that until the ceiling was fixed according to section 86 of the Act, the ownership right vested in the landowner himself and that gift-deed executions under section 84 were legally valid. To this effect it issued a letter to the Taluk Land Boards on 9 November 1973. The ruling of the Kerala High Court on a writ petition on 5 November 1974 (that all partitions and gifts executed after 1 January 1970 by persons holding excess lands were invalid) made such transfers during 1970–74 invalid. But the cheer which this ruling might have brought to the potential beneficiaries of the surplus land redistribution was short-lived.

For, from the time of the court order, implementation of the ceiling provisions in Kerala turned out to be more and more a matter of political expediency. The ruling front, with a view to avoiding the collapse of its fragile facade of unity, yielded to the unreasonable pressures of the leaders of the Muslim and Christian communities, and despite stiff opposition from within and outside the assembly, and despite a statewide land-grab agitation by the CPI(M) for over three weeks in January 1979, gave shape to what has come to be known as the 'Gift Deeds Bill'. This was intended to overcome the hurdle of the court order by legislative means.

The argument of the leaders of these communities was that their personal law did not permit inheritance of parents' property by children. The government's argument was similar. That is, while persons governed by personal law recognising the joint family system which existed on 1 January 1970 got a right to property by birth and the land reforms Act allowed a member of such a joint family to hold lands up to the ceiling limit even after this date as if a partition had been effected, children belonging to families not governed by the joint family system did not acquire any such right as there was no provision in the Act to allow them any share, and the amendment Bill was to put an end to this discrimination. This, however, was not a material argument when the majority of the people in these two communities also had land well below the ceiling, or did not have any land at all for inheritance. Even in the communities which allowed such inheritance, it did not make any significant difference for want of land. The demand for the Bill was obviously prompted by the desire to feather the nests of a few rich landowners who had cornered much of the community's land in their names, and who wanted to retain the same by using the Gift Deeds Bill.

It was during the Emergency that the demand for the Bill began gathering momentum, when the ruling front used it as a bait to hook the Kerala Congress back into its fold. The bait was tempting and the 'help us to help you' attitude of the front towards the Kerala Congress was symbiotic. The Kerala Congress lost no time in joining the front, and after that the front lost no time in working on the Bill. The issue was first discussed by the Liaison Committee of the ruling parties on 15 October 1976. An ordinance in 1978 validating all the gift-deeds executed during 1970–74, and the legislation of the Bill in November 1979 as the Kerala Land Reforms (Amendment) Bill of 1979 replacing this ordinance were the outcome.[19]

The legislative process of the Bill was completed in record time against the heaviest odds. After it was passed by the assembly on 26 October 1979 it was processed in a great hurry for the President's assent before the expiry of the ordinance. It was despatched to Delhi through a 'special messenger' from Trivandrum and officers were drafted to monitor its movement between the various ministries at the Centre before it was put up for the President's signature. Its otherwise prolonged and cumbersome scrutiny by these ministries was short-circuited to process it post-haste. Finally, it was approved by the Centre without any compunction, after having rejected it in February 1979 on a very valid ground that it would in effect reduce the surplus lands and defeat the

[19] For a discussion of the politics of the Gift Deeds Bill, see reports in *Mathrubhumi*, Daily, 16 December 1978, and 12, 19 and 23 January and 12 April 1979. For a discussion of the quick drain of surplus lands and also of the evil effects of Kerala land reforms, see articles in the same Daily, by P. Rajan, on 13, 14, 18 and 20 October 1979. The failure of the ceiling scheme has been discussed by other writers also. See, for instance, Herring, 1980; Krishnaji, 1979; Tharakan, 1982.

very purpose of land reforms. All these clearly showed how powerful lobbies were active in both the central and state capitals.[20]

The enactment of the Bill was undoubtedly a defeat for the CPI(M) and other progressive forces in Kerala, as much as it was a victory for the Kerala Congress and the Muslim League. The real losers were, however, the thousands of landless labourers, especially the Harijans and persons from other socially and economically backward sections of society who were all along hoping to get a piece of land as their own through the surplus land redistribution. The Act legalised the numerous gift deeds executed during 1970–74 that were held invalid by the Kerala High Court, and gift deeds accounted for a major portion of the surplus lands whose take-over was stayed by the court. The area involved in gift deeds alone was estimated as 14,164 acres (see Herring, 1981:66) or about 9 per cent of the area ordered for surrender. Given the average area already distributed as surplus, the number of potential beneficiaries deprived by this Act would have been about 24,007.

## Impact

As already seen, the implementation of the KLRAA resulted in the estimated transfer of 19.7 lakh acres of land to 12.7 lakh households under the first scheme, 0.2 lakh acres to 2.7 lakh households under the second, and 0.5 lakh acres to 0.9 lakh households under the third. While these transfers are expected to be of far-reaching consequences for the agrarian society in Kerala, it is only too obvious that the impact of the first scheme should have been incomparably greater than that of the other two.

### *Relative*

This impact is quite evident from the changes in the land ownership pattern (more of this shortly), the legal abolition

[20] For details, see *Malayala Manorama*, Daily, 27 October 1979; and *Sunday Standard* (Cochin), 4 November 1979.

of traditional landlordism, tenancy, and share-cropping, the emancipation of the tenants from their socio-economic subservience to the landlords, the snapping of the centuries-old nexus between two major agrarian classes, and the supposed elimination of a major constraint on agricultural production. The last point should be seen against the fact that belonging as they did to the high-status castes, landlords in Kerala traditionally spurned labour, remained the leisure-class of society, and were not inclined to either invest in their lands themselves or motivate their tenants to do so.

The impact of the second scheme could be viewed differently. On the one hand, it appears significant when the present land ownership of the *kudikidappukar* is compared, as Herring has done, with their landlessness and the resultant socio-economic subservience in the pre-reforms period, or with the situation of landless labourers elsewhere in India or in the subcontinent, or when it is considered in the context of the exhorbitant price of residential plots in Kerala (see Herring, 1980:66). On the other hand, it pales into insignificance if the extent of land received by a *kudikidappukaran* is compared with that by a cultivating tenant, or if it is borne in mind that the *kudikidappukar* still belong to the large category of agricultural labourers.

In this connection Mencher has observed that in Tamil Nadu the *cheri* (Harijan settlement) was always the property of the Harijans from which the landlord could not evict them at all, and that 'what the legislation in Kerala gave to the landless labourers was something that the Tamil agricultural labourer had all along' (Mencher, 1980:1789). Her views are, however, very formalistic. For one thing, *cheri* by its very notion is a crude expression of an extreme form of social oppression, and the exclusion of which the Harijans in Tamil Nadu are still hapless victims. For another, the homesteads which the *kudikidappukar* obtained under the Kerala Act symbolise their growing socio-political freedom and awareness, thanks to the strong left movement and the weakening of the caste system in the state.

No doubt, the tiny plot of land received per *kudikidappukaran* is too inadequate for agricultural use and a family can hardly subsist on it. The third scheme was intended to supplement this inadequacy, though because of its poor performance it has not achieved this objective. Even its limited impact has been in the Malabar region and not in Travancore where the *kudikidappu* households are concentrated, so that the benefit of these households from the surplus land distribution has been minimal.

### *Aggregate*

The aggregate impact of the land transfers on the agrarian society in Kerala has been considerably impaired by the successive exemption of plantations from ceiling, leading to their continuous strengthening and growing importance in Kerala's economy and polity. Despite this limitation, and despite the limitations of the second and third schemes, the overall impact of the three schemes is still massive and structurally very significant. This is clearly reflected in the post-reforms land ownership pattern, marked by a significant increase in the land-owning households and decrease in the wide inequalities in land ownership. Both these are evident from the data presented in Table 5.9.

As may be seen from the Table, during 1966–67 (that is, on the eve of the implementation of the KLRAA) the land owning households accounted for only about 43 per cent of the total households in the state, whereas in 1982 these accounted for as much as 86 per cent. This increase has been mainly because of the drastic reduction in the landless households from about 57 per cent in 1966–67 to about 14 per cent in 1982. In absolute terms, the reduction is by about 4 lakhs, from about 20 lakhs in 1966–67 to about 6 lakhs in 1982.

One of the reasons for this reduction has been the land transfers under the second and third schemes benefiting more than 3 lakh households, mainly those of agricultural labourers. These transfers also explain the reduction of the households of landless agricultural labourers both in the total households of agricultural labourers from about 30 per cent in 1964–65 to about 16 per cent in 1982, and in the total rural households from about 20 per cent to just 5 per cent during

**Table 5.9** ***Pre- and Post-Reforms Land Ownership Patterns***

| Sl. No | Size Class of Ownership Holding | 1966–67 | | 1982 | | % Change | |
|---|---|---|---|---|---|---|---|
| | | Households | Area | Households | Area | Households | Area |
| 1 | 0.01– 0.99 | 903 | 348 | 2,730 | 753 | 202.3 | 116.4 |
| | | 60.0 | 10.2 | 74.6 | 21.6 | | |
| 2 | 1.00– 2.49 | 337 | 510 | 567 | 861 | 68.2 | 68.8 |
| | | 22.4 | 15.0 | 15.5 | 24.7 | | |
| 3 | 2.50– 4.99 | 135 | 481 | 237 | 812 | 75.6 | 68.8 |
| | | 9.0 | 14.2 | 6.5 | 23.3 | | |
| 4 | 5.00– 9.99 | 88 | 602 | 96 | 644 | 9.1 | 7.0 |
| | | 5.9 | 17.7 | 2.6 | 18.5 | | |
| 5 | 10.00–14.99 | 23 | 272 | 20 | 233 | −13.0 | −14.3 |
| | | 1.5 | 8.0 | 0.5 | 6.7 | | |
| 6 | 15.00–19.99 | 5 | 86 | 6 | 110 | 20.0 | 27.9 |
| | | 0.3 | 2.5 | 0.2 | 3.2 | | |

**Table 5.9** ***Contd.***

| Sl. No. | Size Class of Ownership Holding | 1966–67 | | 1982 | | % Change | |
|---|---|---|---|---|---|---|---|
| | | Households | Area | Households | Area | Households | Area |
| 7 | 20.00–24.99 | 3 | 76 | 1 | 12 | −66.7 | −84.2 |
| | | 0.2 | 2.2 | 0.0 | 0.3 | | |
| 8 | 25.00 & above | 10 | 1,021 | 2 | 54 | −80.0 | −94.7 |
| | | 0.7 | 30.1 | 0.1 | 1.6 | | |
| | Total | 1,504 | 3,396 | 3,659 | 3,479 | 143.3 | 2.4 |
| | | 100.0 | 100.0 | 100.0 | 100.0 | | |
| | Landless Households | 2,027 | | 579 | | −71.4 | |
| | | 57.4 | | 13.7 | | | |
| | Land Owning Households | 1,504 | | 3,659 | | 143.3 | 2.4 |
| | | 42.6 | | 86.3 | | | |
| | Total | 3,531 | | 4,238 | | | |
| | | 100.0 | | 100.0 | | | |

**Note:** Households in number ('000) and percentage area in acres and percentage.
**Source:** Tabulated from Government of Kerala, 1968:213; Government of India, 1986:49, 250.

the same years (see Government of India, 1981, Vol. 1:30; 1986:77). However, their benefit has been only nominal as about three-fourths of their land-owning households still own only less than half an acre of land.

Another, and the most important reason, has been the extensive land transfers necessitated by the first scheme. For, of about 13 lakh tenant households in the state in 1966–67, nearly 10 lakh had not owned any land at all and were depending entirely on leased-in lands (Table 5.10). The implementation of the scheme would have enabled virtually all these households to acquire ownership rights, and the remaining 3 lakh households to augment their holdings.

The redistribution effects of the KLRAA, as reflected in a reduction in the inequalities of land ownership (especially the extreme concentration of lands in a few hands) are evident from the decrease of households and area in the ownership categories of 10 to 15 acres and 20 acres and above, and from the increase of those in the ownership categories below 10 acres. The decrease is especially spectacular in the case of households owning 20 acres and above, while the increase is so in the case of households owning below 5 acres. As about four-fifths of the land transferred under the first scheme and three-fourths under the third scheme were in Malabar, the redistribution effects would have been greatest on this region.

The aggregate impact, impressive though it is in terms of the diffusion of land ownership and reduction in the wide inequalities of the pre-reforms magnitude, however, obfuscates the differentiation of the peasantry and the emergence of a neo-rich class consequent on the land transfers.

In this connection it needs to be mentioned that of the 13 lakh tenant households in the state who were to gain from the first scheme, more than half (56 per cent) held only less than 1 acre and only about one-tenth (13 per cent) of the total leased-in area. Virtually all of them (91 per cent) had no land of their own and were entirely dependent on leased-in lands. But the average leased-in area was only about one-third of an acre (36 cents). The gain of these households would have been only marginal, though certainly not insignificant compared to their pre-reforms status. About a third of the households (35 per cent) held between 1 acre and 5 acres,

Table 5.10 *Tenant Households Depending Entirely on Leased-in Land, 1966–67*

| *Size Class of Operational Holding* | *Travancore* | | *Cochin* | | *Malabar* | | *Total* | |
|---|---|---|---|---|---|---|---|---|
| | *Households* | *Area* | *Households* | *Area* | *Households* | *Area* | *Households* | *Area* |
| 0.01– 0.99 | 38 | 13 | 155 | 48 | 462 | 180 | 655 | 241 |
| | 79.2 | 34.2 | 76.0 | 21.9 | 63.8 | 20.9 | 67.1 | 21.5 |
| 1.00– 4.99 | 9 | 18 | 39 | 81 | 238 | 477 | 286 | 576 |
| | 18.7 | 47.4 | 19.1 | 37.0 | 32.9 | 55.3 | 29.3 | 51.5 |
| 5.00–14.99 | 1 | 7 | 9 | 61 | 23 | 194 | 33 | 262 |
| | 2.1 | 18.4 | 4.1 | 27.9 | 3.2 | 22.5 | 3.4 | 23.4 |
| 15.00 & above | — | — | 1 | 29 | 1 | 11 | 2 | 40 |
| | | | 0.5 | 13.2 | 0.1 | 1.3 | 0.2 | 3.6 |
| Total | 48 | 38 | 204 | 219 | 724 | 862 | 976 | 1,119 |
| | 100.0 | 100.0 | 100.0 | 100.0 | 100.0 | 100.0 | 100.0 | 100.0 |
| | 4.9 | 3.4 | 20.9 | 19.6 | 74.3 | 77.0 | 100.0 | 100.0 |

**Note:** Dash indicates nil. Households in number ('000) and percentage; area in acres and percentage.
**Source:** Tabulated from Government of Kerala, 1968:171–2.

and two-fifths (42 per cent) of the total leased-in area, with about 1.8 acres as the average size. Their gain, though not substantial, was not inconsiderable. Though only a tenth (9 per cent) of the households held more than 5 acres, their holdings accounted for more than two-fifths (44 per cent) of the total leased-in area, with about 7.3 acres as the average size (Table 5.11). It is these households who would have gained substantially, especially those holding more than 15 acres.[21]

### *Emerging Agrarian Situation*

In understanding the differentiations resulting from these differential gains, it may be said, at the risk of generalisation, that in the non-plantation agriculture there are now at least three distinct groups in Kerala as a whole. At one extreme is a small group of farmers who gained substantially from the land transfers. Many of them do not till their lands themselves but only supervise the work of hired labourers. A neo-rich class within these farmers seem to be fast emerging as socially and politically the most important group in the non-plantation sector.

[21] For a discussion of the differential impact of the first scheme on the tenantry, see Raj and Tharakan, 1984:59–62; also, Herring, 1983:211–12.

While on the differential gains, the differential loss could also be noted. According to the LRS data, more than half of the landlords (54 per cent) had owned less than 5 acres of land, about one-fourth (24 per cent) between 5 and 15 acres, and the rest (22 per cent) 15 acres and above. The first two classes had leased out nearly half (46 per cent) and the third about two-thirds (67 per cent) of their lands (see Government of Kerala, 1966:154, 169). Of these three classes the first and the second would have felt the impact of their loss more severely. In the context of Malabar, the Nambutiris seem to have been the worst affected. Once constituting the aristocracy, many of them now find it difficult to get any alternative source of income. In the pre-reforms period, for the poorest of the poor of them priesthood was an adequate and alternative source of income. But, as many of the temples which were feeding them in the past have also fallen into disuse, the income derived from this profession now is too low to sustain a family. No doubt, in any structural transformation in society some are bound to suffer. As such, the hardship of the former landlords is no justification to see the impact of the scheme in a negative light. For reports on the decadence of some of the important temples in Malabar consequent on the implementation of the KLRAA, see *Mathrubhumi*, Daily, 11 and 18 June 1979.

**Table 5.11** *Tenant Households by Leased-in Lands, 1966–67*

| *Size Class of Operational Holding* | *Travancore* | | *Cochin* | | *Malabar* | | *Total* | |
|---|---|---|---|---|---|---|---|---|
| | *Households* | *Area* | *Households* | *Area* | *Households* | *Area* | *Households* | *Area* |
| 0.01– 0.99 | 75 | 21 | 172 | 53 | 474 | 184 | 720 | 259 |
| | 46.9 | 7.7 | 56.6 | 11.2 | 57.7 | 15.7 | 56.1 | 13.5 |
| 1.00– 4.99 | 60 | 78 | 94 | 166 | 293 | 567 | 447 | 811 |
| | 37.5 | 28.6 | 30.9 | 35.1 | 35.7 | 48.3 | 34.8 | 42.2 |
| 5.00–14.99 | 19 | 67 | 32 | 153 | 48 | 313 | 99 | 533 |
| | 11.9 | 24.5 | 10.5 | 32.3 | 5.8 | 26.6 | 7.7 | 27.7 |
| 15.00 & above | 6 | 107 | 6 | 101 | 6 | 111 | 18 | 319 |
| | 3.7 | 39.2 | 2.0 | 21.4 | 0.7 | 9.4 | 1.4 | 16.6 |
| Total | 160 | 273 | 304 | 473 | 821 | 1,175 | 1,284 | 1,922 |
| | 100.0 | 100.0 | 100.0 | 100.0 | 100.0 | 100.0 | 100.0 | 100.0 |

**Note:** Table includes 99,708 acres of land leased in by 16,251 landlord households. These households were, however, excluded by the LRS from its estimate of tenant households. Households in number ('000) and percentage; area in acres and percentage.

**Source:** Tabulated from Government of Kerala, 1968: 185–202.

At the other extreme is a large group of agricultural labourers, for many of whom the only benefit out of the land transfers has been the tiny plot of homestead land. For their livelihood they still sell their own labour and depend on the groups above them in the agrarian hierarchy. But there seems to be a progressive reduction in the chances of their getting employment on the farms, as the share of the physical labour contributed by the post-reforms small and marginal farmers seems to be on the increase. Thus the speeding up of the pauperisation of the rural masses seems to be one of the unwelcome consequences of the implementation of the present Act.

In between these two groups is a large group of marginal and small farmers, who were marginal and small tenants in the pre-reforms period. If their holding is too small as not requiring any wage labour, then it will not be viable and they have to sell their own labour for their living as it is being done now. If the holding is reasonably big, then to make it viable its owner has to have his own cattle and the required number of men and women at home ready for work. But this seldom happens and he has to depend on wage labour. However, employing wage labour has not at all been profitable for many of the marginal owners; as agricultural labourers have become an organised force demanding more and more wages from the landowners. The recent emergence of share-cropping in different parts of the state, partly because of the provision in the Gift Deeds Act for validating certain tendencies created after 1964, indicates the inability of these farmers to stick to their holdings, and probably the inevitability of tenancy.

These and the ability of the rich farmers to corner the gains of the production and marketing processes and of the governmental measures, seem to indicate the emergence of a more parasitic agrarian situation in this state of 'model land reforms'. It is in this context that the next phase of land reforms assumes great significance. But considering that Kerala is perhaps the only state to implement radical land reforms whereas the Centre and virtually all the states have already buried deep the very idea of land reforms, the left in the state may have to wait for a longer time to complete its mission. That is also a dilemma of its uneven development in the country.

Meanwhile the immediate impact of the reforms on the land system and its consequences for the agrarian society in the Malabar region will be examined in the following chapters in the context of a selected village.

# 6

# Land Reforms and Changes in Land System, 1970–79

Kodakkad, the village selected for field investigation, is one of the earliest settlements in Kerala.[1] It is the southern-most village of old Kasargod taluk in the erstwhile British administrative division of south Kanara. Karivellur on its south and Peralam on its east in the present Cannanore district are the northern-most villages of erstwhile British Malabar. It is about 6 kms east of the Bombay-Comorin national highway. At a distance of about 5 kms northwest of it is the nearest railway station of Chervathur, and about 14 kms southwest of it is the nearest town of Payyanur.

The village society in the pre-reforms period,[2] understood

[1] This is revealed by megaliths unearthed from this village in 1969 by K.K.N. Kurup, as mentioned by him in a personal communication.

[2] Though the reforms under reference is the 1969 Act, the pre-reforms period should not be narrowly construed as what immediately preceded the beginning of the implementation of this Act. For one thing, in understanding social phenomena, especially the process of social change, it is meaningless to apply a narrow cut-off. For another, as mentioned earlier, in the agrarian society of Kerala the period 1957 to 1969 was an interface between the initiation and implementation of this Act. Accordingly, the reference to the pre-reforms period here is in a very broad sense and to different periods

in terms of its caste system and caste-based land system, had been a fair specimen of the larger society in Malabar; its peasantry had been active participants in the third and most active phase of the peasant struggles in the region; and its record of the implementation of the 1969 Act closely conforms to the larger state pattern. Although caste has already lost much of its oppressive and exploitative virulence in the village, to all intents and purposes it is still very important in the village society. This, the correspondence between caste and class in the traditional village society, and the fact that the caste system legitimised and reinforced the iniquitous land system of the pre-reforms times, explain the rationale of the detailed caste-analysis of the village study.

## Caste System

The village represents all the major castes of Malabar. Following the discussion in Chapter 2 and following Innes' classification of the castes in Malabar, (1951:105–37) its various castes are classified into five hierarchical groups as shown in Table 6.1.[3] Some of the caste appellations in the Table are derogatory by the contemporary societal norms. However, it is difficult to do away with them because of the need to understand the principal social groups as historical communities.

The Nambutiris mentioned in the Table are both the Brahmins of the Parasurama tradition, and the Embrantiris.[4]

---

of time depending on the nature of the background information relevant to an understanding of the impact of the 1969 Act.

[3] At the time of field investigation the total number of households in the village was 1,162 (according to the panchayat office estimate), of which data were collected from 800 households. For a note on the village and the village survey, see the appendix to this chapter.

[4] Brahmins of the Parasurama tradition are the Nambutiris proper or the Brahmins believed to have been brought from the north by Parasurama. Embrantiris are the Brahmins believed to have been brought from Gokarnam by one of the early rulers of Chirakkal following the Parasurama tradition. According to local lore they were brought in two batches of 257 and 237 families and settled in five *gramams* and four *gramams* respectively, and are also known as *anchudesakkar* (people of the five *desams*) and *naludesakkar* (people of the four *desams*).

**Table 6.1** ***Households by Caste/Community***
***(Households in Number and Percentage)***

| *Sl. No.* | *Caste/Community* | *Number* | *Percentage* |
|---|---|---|---|
| 1 | Nambutiris | 75 | 9.4 |
| 2 | *Ambalavasis* | | |
| 2.1 | Nambis | 2 | |
| 2.2 | Nambissans | 4 | |
| 2.3 | Variyars | 2 | |
| 2.4 | Marars | 3 | |
| 2.5 | Pothuvals | 21 | 4.0 |
| 3 | High Caste Nayars | 58 | 7.2 |
| 4 | *Low-Caste Nayars* | | |
| 4.1 | Maniyanis | 150 | 18.7 |
| 4.2 | Muvaris | 36 | 4.5 |
| 4.3 | Vaniyas | 75 | 9.4 |
| 4.4 | Veluthedans | 4 | |
| 4.5 | Vannathans | 2 | |
| 4.6 | Navutiyans | 8 | 1.7 |
| 5 | *Higher Polluting Castes* | | |
| 5.1 | Tiyyas | 235 | 29.4 |
| 5.2 | Asaris | 12 | |
| 5.3 | Kollans | 4 | |
| 5.4 | Chettis | 10 | |
| 5.5 | Kavutiyans | 5 | |
| 5.6 | Vannans | 9 | |
| 5.7 | Malayans | 2 | |
| 5.8 | Kanisans | 10 | 6.5 |
| 6 | *Lower Polluting Castes* | | |
| 6.1 | Pulayas | 16 | |
| 6.2 | Mavilons | 7 | |
| 6.3 | Madigas | 19 | |
| 6.4 | Kuravans | 5 | 5.9 |
| 7 | Mappilas | 26 | 3.2 |
| | Total | 800 | 100.0 |

**Source:** Tabulated from the field-data.

As some of the Nambutiri respondents recounted, the Embrantiris of this village as also of several other villages in north Malabar had usurped the title Nambutiri from at least the early nineteenth century. Both these groups are patrilineal. Over the years they have merged into one interdining and intermarrying group. Moreover, they together account for the bulk of the former *janmis* in the village. Accordingly, they are treated here as one group.

In the absence of Pattars and royal lineages, in the caste hierarchy of the village the Ambalavasis ranked immediately below the Nambutiris. Of them some worked as temple functionaries and others as cultivating tenants of the Nambutiris.

The village represents seven of the Nayar castes of north Malabar. Of the high-caste Nayars or Nayars proper (comprising Adiyodis, Kurups, Nambiyars, and Nayars), some were small *janmis*, and the rest were cultivating tenants of the Nambutiris. Of the low-caste Nayars, Maniyanis (also known as Erumans) were cowherds and cultivating tenants; Muvaris[5] were masons and stone-cutters whose women worked as maids in Nambutiri *illams*; Vaniyas, also known in other parts of Malabar as Vattekkadan or Chakkingal (*Chakku*: oil press) Nayars were oil pressers and cultivating tenants; Vannathans and Veluthedans were washermen; Navutiyans, a division of the Vilakkattalavan caste, were barbers of the non-polluting castes.

Of the higher polluting castes, Tiyyas were toddy-tappers, cultivating tenants, and agricultural labourers; Asaris (carpenters), Kollans (ironsmiths), and Chettis (goldsmiths)[6] are artisans; Vannans, a matrilineal caste of north Malabar and the counterparts of the patrilineal Mannans of south Malabar were washermen of the higher polluting castes, Teyyam-dancers,[7] and specialists in the preparation and application

[5] Maniyanis and Muvaris are immigrants from the Tulu country and belong to the generic category of Uralis. See Innes, 1951:121.

[6] Chettis of the village are also called Pon-Chettis (*pon*:gold). They are treated here as an artisan caste corresponding to the Thattans of other parts of Kerala.

[7] Teyyam is an indigenous institution of ritual dances. The term also denotes any of the numerous dances within it.

of ayurvedic medicines; Malayans, a patrilineal caste of north Malabar and the counterparts of the Panans in south Malabar were musicians, exorcists, and Teyyam-dancers; Kavutiyans were barbers of the higher polluting castes; Kanisans were astrologers.

Of the lower polluting castes, Pulayas were agrestic serfs attached to the Nambutiri and Nayar families. Mavilons, a small tribe found in the northern part of north Malabar, made baskets, mats, and fences, and collected honey and other forest produces. They were also Teyyam-dancers. Madigas (also known as Chakkiliyans), an immigrant group from Karnataka, worked as cobblers, and prepared lime for white-washing. Kuravans were gypsies. The village has only a few Mappilas, who as a fringe-group of petty traders had hardly any role in the village land system and social order.

For analytical purposes the various subdivisions of the Ambalavasis are treated here as one group; three of the low-caste Nayar groups—namely, Veluthedans, Vannathans, and Navutiyans—and all but the Tiyyas of the higher-polluting castes as one group of artisan and service castes; and all the lower polluting castes as one group of Harijans.

## Land System

According to the revenue records of the village office, the village is 4,161 acres in area. Of this 1,019 acres (24.5 per cent) are mentioned as unoccupied dry land, 888 acres (21.3 per cent) as occupied dry land, 65 acres (1.6 per cent) as 'poromboke,' 1,719 acres (41.3 per cent) as garden land, and 470 acres (11.3 per cent) as wet land. Of these land types, unoccupied dry land consists of forest land and land reserved for public purposes; 'poromboke' is land used for public roads, and the remaining three types are private lands. Of the private lands, garden and wet lands are hereditarily owned by the villagers, while occupied dry land is waste land assigned to them by the government.[8]

The above classification, however, has been made mainly

[8] The large area of government land, mainly unoccupied dry land, has to be explained by the difference between the British revenue administration in south Kanara and Malabar. While the revenue settlements in both regions

for purposes of revenue collection, and there is not much qualitative difference between garden and dry lands. Hence, private lands in the village are classified here into two broad types of wet and garden lands, and dry land is included in the latter.

Of these two types of private lands, although the extent of garden land is relatively large, in the pre-reforms period it was neither an economic resource nor a source of social esteem. For, it was mostly the area occupying a person's dwelling, the compound around it and the assessed waste, and its use for cultivation that presupposed the possession of resources which many of the villagers lacked. Even if some villagers did not lack the resources, they were reluctant to make any investment as it could not yield any immediate returns. Hence, the garden land was mostly left free for the natural growth of timber and fruit trees.

Although the extent of wet land is too scarce, that alone is used for the cultivation of paddy, the staple crop of the village, and usually two, and sometimes even three crops are grown. Hence a lot of value is attached to its ownership and control. In the pre-reforms period there were instances of villagers wangling an acre or two of wet land for leasing out to a few tenants just for acquiring the honorific title 'Janmi'. Among the tenants, there was always a scramble to get some wet land on lease. As persons who lost wet land as either *janmis* or tenants were looked down upon, right in wet land was jealously guarded and seldom alienated. It is against this background of the relative scarcity and greater importance of wet land that the pre-reforms land system has to be understood.

In this land system the *janmis* were mostly the Nambutiris and partly the Nayars, the various categories of tenants were mostly the Ambalavasis, Nayars, Maniyanis, Vaniyas, and the Tiyyas, and the agricultural labourers were mostly the Tiyyas and the Harijans, and partly the Maniyanis and the Vaniyas.

---

were ryotwari with some variations and zamindari implications, in Malabar the entire land belonged to the *janmis* who had absolute ownership rights. In south Kanara district, however, government was the absolute owner and the *janmi's* right was only derived from it.

## Land Tenures

The pattern of land tenures in the village was broadly the same as in the larger society. But, the different tenures and the terms used for them varied widely, and probably reflected such variations between south and north Malabar as reported by Logan and as might have taken place in the subsequent years. In the course of field investigation five types of tenures were identified in the village, three on wet land and two on garden and dry lands.

Of the tenures on wet land, the first was on oral lease (*vakkal kaivassam*). It involved neither any written agreement between the *janmi* and the tenant nor any payment of advance by the latter, but only an implicit agreement on the part of the tenant to pay the stipulated rent and customary dues promptly. The tenurial right was non-transferable. Since the *janmi* could terminate it at will, there was no general duration of tenure either. The second tenure was differently known as *verumpattam*, *verumkhozu*, or registered *maru-pattam*. It involved both written agreement and payment of advance, normally of one year's rent. Though the duration of the tenure was specified, more often than not the tenant held the land beyond the specified time. The tenurial right was transferable by the tenant subject to the compliance of the conditions of the original agreement by the subsequent tenant. The third tenure was *panayam*, a mortgage for long periods. Until the mortgage money (also called *kanam*) was repaid, the mortgagee could use the land as his own without paying any rent or customary dues to the *janmi*. The *janmi* mentioned in the agreement the reasons for the mortgage. This mortgage had acquired the nature of a tenure (called *kanam-kuzhikanam*) and was deemed so by the 1969 Act.

Of the tenures on garden and dry lands, the first was *kuzhikanam*, a lease usually for twelve years. While some *janmis* leased out the land without collecting any advance, some took a small sum, called, *manusham*. The tenure was on the basis of either a written or a tacit agreement. The rent, called, *purappad* or *pattam*, was collected in either cash, or kind (in the form of usufructs). The tenant had no right to cut the trees without the consent of the *janmi*, though he could improve the land. The second tenure was *otti*, a usufructory mortgage in which the *janmi* merely retained the property title and the right to redeem.

While the pattern of tenancy thus clearly reflected the virtual dependence of certain social groups on the Nambutiris because of the latter's dominance in the village land system, certain other social groups were reduced to even greater dependence as agricultural labourers, homestead tenants, household servants, and so on. Such dependence in turn had led to the social subservience of all the dependent groups which was perpetrated by the Nambutiris by insisting on strict adherence to traditional caste-practices and caste-based socio-economic relations. Coupled with this dependence was the uncertainty in the possession of land by the dependent households, arising out of the anomalies in the nature of land revenue collection by the government. Though land revenue was settled directly with the *janmis* in both south Kanara and Malabar, it was the tenants who paid it to the government subject to later adjustment by the *janmis* in the rent payable to them. Consequently, more often than not the tenants themselves had to bear the entire burden of government's revenue demands on *janmi* lands, including on the assigned and uncultivated waste. Failure in revenue payment led to, first, attachment of the standing crops, and then, sale of the lands through auction. Though the action was directed against the *janmis*, it was the tenants who suffered most, by losing their crops and even their leased-in holdings. It is in the context of this intricate nexus between land and society enmeshing the landless and the land-poor in a servile social trap that the participation of the village peasantry in the numerous peasant struggles of the region and the implementation of the 1969 Act have to be understood as vital forces of social change.

## Peasant Activism

Though Kodakkad may not find a place in the history of Kerala, to the peasantry of north Malabar its name is unforgettable, thanks to its important role in the third and most active phase of the peasant struggles. This role was, however, not confined to the village, but, in keeping with the broad regional character of the peasant struggles, transcended the village boundaries and took the nature of reinforcing those larger struggles of the region as already discussed in the preceding chapters. As this makes it rather difficult to understand peasant activism in the village *per se*, the discussion in this section will be sketchy and, at best, a chronicle of the main events still recounted by the peasant leaders of the region.

As the Mappila outbreaks and the Nayar agitations were virtually confined to south Malabar, it is only since 1934, when the peasant struggles began to strike roots in north Malabar, that the village peasantry began to be drawn into peasant activism. In April that year, a youth organisation for social reforms—the *Abhinava Bharat Yuvak Sangham* (ABYS)—was formed at Karivellur by A.V. Kunhambu, M.P. Appu, P. Kunhiraman, and a few others. Its activities embraced the villages of Karivellur, Kodakkad, Peralam and Vellur. Koyyan Kunhikannan, Elichi Kannan, T.V. Sankaran Master, Paringen Kunhiraman Nambiar, Puliruveetil Krishnan and others were active in promoting it in Kodakkad and other villages in the Kasargod taluk. On the initiative of this organisation, in September the same year the *Karivellur Samyukta Karshaka Sangham*—the first regular unit of the *Karshaka Sangham* (KS) covering the same villages—was formed at Karivellur, with A.V. Kunhambu as president, and M.P. Appu as secretary.

Against this organisational background, in 1935 a special peasant conference was organised at Karivellur, presided over by A.K. Gopalan. This was the first peasant conference to be attended by the peasantry of Kodakkad and other villages of the Kasargod taluk. Within the next three years units of the Kerala Pradesh Congress Committee (KPCC) and the KS were organised in the numerous villages of Kasargod taluk. These units enlisted a large number of peasants in the Congress Party and the KS, and united them in the fight against the British and the *janmis*.

At the first conference of the Kasargod taluk peasantry organised at Kanhangad in 1937, presided over by K.P.R.

Gopalan, the *Kasargod Taluk Karshaka Sangham* (KTKS) was formed, mainly on the initiative of K.A. Keraleeyan, A.V. Kunhambu, T.S. Tirumumb, V.V. Kunhambu, N.S. Namboodiri, K. Madhavan, and V.M. Vishnu Bharateeyan. They were later joined by several local leaders like Koyyan Kunhikannan, P. Ambu Nair, N. Narayana Varier, T.V. Kunhambu, and T.K. Chandan.

Meanwhile, a joint agitation of the KPCC and the Kerala Congress Socialist Party (KCSP) for merging Kasargod with Malabar led to a conference at Nileswar in April 1938, presided over by Muhammed Abdur Rahman Sahib, an eminent KPCC leader and inaugurated by Kongattil Raman Menon. Highlighting the tenancy problems in the Kasargod taluk, the conference passed a resolution demanding permanent occupancy rights to the tenants in the region. In December that year, as part of the all-Malabar peasant rally, a *jatha* of 130 persons left from Olavara in south Trikaripur to present a memorandum to the district collector at Mangalore. It was led by T.S. Tirumumb, president of the KTKS and seen off by A.V. Kunhambu, and K.A. Keraleeyan, leaders of the *Malabar Karshaka Sangham*. Koyyan Kunhikannan was its treasurer. The memorandum demanded prohibition of illegal exactions, extension of the proposed Malabar tenancy legislation to Kasargod taluk, and the merger of the taluk with the Malabar district.

The organised strength of the peasantry during this period as discussed in Chapter 3 was clearly reflected in the village as well, and as in the larger society, the KS drew its strength and sustenance mainly from the lower peasantry especially from the Tiyya caste, though members of the other castes of tenants and agricultural labourers also supported and cooperated with its activities. One instance of this organised strength was an interesting social boycott. When the *potail* or village headman, an upper-caste *janmi* slapped the lower-caste *kayattukaran*[9] on the ground that the latter failed to offer him the customary feudal levies, all the villagers save the *janmis* turned against him withholding the customary services. The headman (who could neither help himself to the use of a razor nor get the

[9] Literally, rope-bearer: person protecting the fields from cattle.

services of a barber) was forced to apologise publicly to the *kayattukaran* in order to get the services restored. Boycotting the *polu*[10] of the biggest *janmi* in the village was another instance. Ordinarily the peasants had to cringe before this *janmi* and do his mean errands to get a *polu*. As the KS had declared this *janmi* as the biggest enemy of the peasantry, some of its well-to-do and first-ranked members like Koyyan Kunhikannan, Karikutty, and Kaniyan Krishnan decided to teach him a lesson by boycotting his *polu*. For this purpose they pooled their resources and, with the help of the KS, built up a reserve of paddy for distribution among the needy members of the KS.

The fourth annual conference of the ABYS and the second annual conference of the KTKS held at Kodakkad in January 1939 are perhaps the most memorable events in the peasant history of the village. Both were well-organised mammoth conferences attended by about 10,000 persons. Each had a separate reception committee, and a separate president and secretary. The president and secretary of the youth conference were E.V. Narayanan Nambi and T.V. Sankaran Master respectively. Those of the peasant conference were T.S. Tirumumb and P.C. Kunhiraman Adiyodi. Koyyan Kunhikannan was treasurer of both. The youth conference was held on 14 January, presided over by A. Kameswar Rao, and inaugurated by Muhammed Abdur Rahman Sahib. The peasant conference was held on 15 January, presided over by N.G. Ranga, general secretary of the AIKS, inaugurated by Moidu Moulavi, an eminent member of the KPCC, and participated by several peasant leaders of Kerala such as A.K. Gopalan, P. Krishna Pilla, K.A. Keraleeyan, P. Narayanan Nair, and K.P.R. Gopalan. It was marked by inter-dining and the staging of a popular drama, *Pattabakki* (Rent-Arrears), and also by a women's conference participated by women from several peasant and workers families in the area.

Yet another event remembered for the involvement of the village peasantry is the Kayyur riot in the neighbouring village. The involvement took place in several ways. The village in

[10] A usurious paddy-loan at the rate of three *paras* in return for ten *paras*.

general was a hideout for the accused and a hunting-ground for the police, with the police-force camping at the *illam* of the biggest *janmi* of the village; some of the village peasantry were participants in the events which preceded the death of the police constable including the *jathas* which confronted him; finally, three of the non-participants from the village were among the accused in the riot. They were Vaniyan Sankaran Master (secretary, *Kodakkad Karshaka Sangham*, from 1935 to 1940), Koyyan Kunhikannan (treasurer, KTKS from 1938 to 1939; president, Kodakkad Congress Committee, and member, Kasargod Taluk Congress Committee), and Elichi Kannan—all marked by the British administration for their active role in the peasant struggles of the region.

The village peasantry were participants in the 1946 agitation against restrictions on the collection of fodder and firewood from the neighbouring Cheemeni estate—a cashew estate of about 6,000 acres of a later MP of the Congress Party, leased in from two of the biggest *janmis* of the region, namely, the *tazhekkat mana* and *kuttamath kunniyur* house, the former a Nambutiri *illam*, and the latter a Nayar *taravad*. They were also participants in the Karivellur agitation in the same year against the lifting of paddy by the Chirakkal Raja. In this agitation two persons were shot dead by the police and A.V. Kunhambu, one of the peasant leaders of the region who spearheaded it, was imprisoned.[11] In 1948, following the Calcutta thesis, and as part of a larger agitation, the KS organised a *jatha* to the biggest *janmi* of the village,[12] for the

[11] For a brief account of peasant activism in the Kasargod taluk including in Kodakkad, see, Kurup, 1978:9–33.

[12] In the early 1960s, in a frenzied mood this *janmi*, notorious for his tyranny and oppression, shot dead a Nayar villager in a land-dispute. Koyyan Kunhikannan, who left the peasant and communist movements after his punishment as an accused in the Kayyur riot had a highly emotional story to tell about the cruelty of this *janmi*. The *janmi* was after his blood for a long time as he was a peasant activist. Once the *janmi* frantically chased him through the village with a rifle in his hand. When Koyyan became helpless, by sheer providence he ran into the house of a poor villager, a Chetti, where there was none but a child lying in his cradle. He took the child in his arms and faced his enemy knowing full well that the latter would not 'commit the sin' of infanticide. That act did really save him from death.

purchase of excess paddy from him for distribution among the villagers.[13]

While the village peasantry played an important role in the peasant struggles of the region in the subsequent years as well, it was not as vital as in the earlier times. This was probably because of the formation of the unified Kerala state, and the new political role assumed by the communist party in initiating land reforms for meeting the long-standing demands of the peasantry. The subsequent role of the peasantry was during the implementation of the 1969 Act, in which the KS took an active interest in educating the peasantry on the provisions of the Act and in helping them file purchase forms before the Tribunals. However, since the Act itself was an exercise in abolishing landlordism and tenancy, and its implementation, already taken as a *fait accompli*, was smooth and did not evoke much protest from the *janmis*, this role was not of much significance.

## Implementation of the 1969 Act

To recapitulate, the three schemes envisaged by the Kerala Land Reforms (Amendment) Act of 1969 were: (*a*) conferment of ownership rights on the cultivating tenants; (*b*) granting option to the *kudikidappukar* to purchase their homesteads; and (*c*) acquisition and distribution of surplus lands.

For the implementation of the first two schemes the village was under the two Land Tribunals at Nileswar, which were in charge of all the 16 villages in the Nileswar Block. For that of the third scheme it was under the Tahsildar, Hosadurg taluk. Data collected from these two sources indicate the progress in implementation of the three schemes.

[13] According to a statement made by T.S. Tirumumb when interrogated at the nearby Chervathur police station on 21 May 1948, he led a *jatha* to Karindalan Koman, appealed to him and obtained from him 10 *paras* of paddy. He paid Rs. 20 as the price of the paddy but the *janmi* refused to take it. Such 'paddy-purchasing' was conducted under his instruction in other places such as Madikai, Kayyur, Kilaikode, and Kodakkad. However, he attended only the Karindalan *jatha*. See, GO 71, Confidential, 29 November 1948, Under Secretary's Safe, Government of Madras.

**Table 6.2** ***Transfer of Ownership Rights as on 30 April 1979***

| *Particulars of Applications* | *Scheme 1* | | *Scheme 2* | |
|---|---|---|---|---|
| | *Number* | *Percentage* | *Number* | *Percentage* |
| Allowed | 41,060 | 76.2 | 603 | 58.9 |
| Rejected | 12,818 | 23.8 | 416 | 40.6 |
| Pending | 32 | 0.1 | 5 | 0.5 |
| Total | 53,910 | 100.0 | 1,024 | 100.0 |

**Note:** Scheme 1: Conferment of ownership rights on cultivating tenants; Scheme 2: purchase of *kudikidappus* by *kudikidappukar*.

Under the third scheme, as on 7 April 1979 a total area of 30,943 acres of land was ordered for surrender. Of this 26,232 acres (85 per cent) were taken over; 4,025 acres (13 per cent) were covered by stays including 154 acres in gifts; and the balance for take over was 686 (2 per cent) acres. Village-wise data are not available to ascertain the relative progress in the implementation of these schemes. But data collected from Kodakkad clearly show the progress within this village in terms of the actual land transfers.

By the implementation of the first scheme, ownership rights of about 162 acres of wet land and 278 acres of garden land were transferred from *janmis* to cultivating tenants. The transfer of wet land benefited about two-thirds of the present wet land owning households, while that of garden land benefited about one-third of the total households. In a few instances the *janmis* had sold some of the leased out holdings to their tenants anticipating the implementation of the Act. However, since the Act itself contained provisions for settling cases on a mutual basis outside the official channels, such anticipatory sales have not defeated its purpose. Moreover, in the course of field investigation it was discovered that all the cultivating tenants in the village had obtained ownership right of the lands leased in by them.

By the implementation of the second scheme, six of the respondent households obtained an ownership right of about ten cents each of homestead land. The small number of beneficiaries does not mean that the scheme was not efficiently implemented. For, on the eve of the implementation of the 1969 Act, the *kudikidappus* in Malabar accounted for only less than one-fourth (23 per cent) of the total *kudikidappus* in the state. In this connection it should also be mentioned that until about a few years ago, the government had distributed at nominal prices a large area of waste land among the landless and land-poor persons in the village. Although such distribution was done to convert waste land into revenue land, it benefited many of the villagers, especially the Harijans. Further, as part of a larger scheme for the settlement of landless agricultural labourers implemented at the instance of the government of India, between 1964 and 1969 about 200 families in the village obtained about 4 acres each of government waste land, and a grant-cum-loan of Rs. 1,000 to Rs. 1,500 for land improvement. These steps considerably reduced the number of homesteads in the village.

By the implementation of the third scheme, six of the respondent households obtained the ownership right of about 1 acre each of garden land. The small number of beneficiaries in this case is a clear indication of the failure of the scheme. Owing to bogus transfers, the Land Board could not take over even one-fifth of the surplus land. In the course of field investigation three cases of such transfers were discovered. In one case a person owning about 50 acres of land had partitioned his land in different names. In the other two cases, one person owning about the same extent and the other owning about 60 acres of land had 'gifted' some of their lands to their married daughters living separately in the same village. The transfers, in all the three cases, were effected just before the 1969 Act was enforced to avoid surrender of the surplus and the land was actually in the possession of these three persons as it was frankly admitted by one of them. It is interesting to note that these surplus land owners were cultivating tenants in the pre-reforms period, two of whom were also frontline peasant activists.

Comparison of the pattern of implementation of the three schemes in the village with that at the state level clearly indicates that the former closely conforms to the latter; that in terms of the total area of land transferred and the number of households benefited by it, the first scheme has been the most important plank of the Act; and that as at the state level it is only the implementation of this scheme which would have had any significant impact on the village land system. To understand this impact clearly, it is first of all necessary to have an assessment of the loss and gain of lands by the different social groups in the village as a result of the implementation of this scheme.

### *Losers and Gainers*

Tables 6.3 and 6.4 show the wet and garden lands lost by the different groups through the Land Tribunals, expressed as percentage of the wet and garden lands owned by them on the eve of the implementation of the Act.[14] As may be seen from Tables 6.3 and 6.4, it is only to the Nambutiris that the loss has been great. Among them, virtually all (92 per cent) the wet land owning households and more than two-thirds (68 per cent) of the garden land owning households of the pre-reforms period have lost land, with virtually all of the former (95 per cent) and nearly half of the latter (45 per cent) losing between 41 per cent and 100 per cent of their holdings. Of the households which have lost wet land, as high as two-thirds (66 per cent) have lost between 61 per cent and 100 per cent of their holdings, whereas of the households which have lost garden land, only one-fifth have lost such a high proportion.

[14] While data on post-reforms land ownership are mainly based on the tax-receipts and in the absence of such receipts, were cross-checked with the Basic Tax Register of the village office, data on pre-reforms land ownership are based mainly on oral information furnished by the respondents. As the area of land owned by some of the *janmis* was so large and fragmented into so many tenanted plots, and as most of them had not kept any accurate record of their holdings, they could not give the actual area of land owned by them. This limitation of data on pre-reforms land ownership also applies to the data on loss of land by the former *janmis*. In a particular instance, the *janmi* had owned land in about half a dozen villages and leased out most of it to hundreds of tenants. He managed the land himself with the help of a *karyasthan* (manager). However, by the time the land reforms Act was implemented the *janmi* died, leaving the entire burden of land management to his teenage son. He was not aware of the area of land owned and leased out by his father, and all that he could offer by way of information was a bundle of notices issued by the Land Tribunals. As the Land Tribunals had not kept details for each village separately, it was not possible to collect information from them either.

**Table 6.3** ***Households by Loss of Wet Land Through Tribunals (Households in Number and Percentage)***

| *Sl. No.* | *Caste/Community* | *Percentage Loss of Land* | | | | | *Total* |
|---|---|---|---|---|---|---|---|
| | | *81 to 100* | *61 to 80* | *41 to 60* | *Up to 40* | *Nil* | |
| 1 | Nambutiris | 28 | 11 | 17 | 3 | 5 | 64 |
| | | 43.8 | 17.2 | 26.6 | 4.7 | 7.8 | 100.0 |
| 2 | Ambalavasis | — | — | — | 2 | 7 | 9 |
| | | | | | 22.2 | 77.8 | 100.0 |
| 3 | Nayars | 5 | 2 | — | 3 | 13 | 23 |
| | | 21.7 | 8.7 | | 13.0 | 56.5 | 100.0 |
| 4 | Maniyanis | — | — | — | — | 32 | 32 |
| | | | | | | 100.0 | 100.0 |
| 5 | Muvaris | — | — | — | — | — | — |
| 6 | Vaniyas | — | — | 1 | 1 | 14 | 16 |
| | | | | 6.3 | 6.3 | 87.5 | 100.0 |
| 7 | Tiyyas | — | — | — | 1 | 49 | 50 |
| | | | | | 2.0 | 98.0 | 100.0 |
| 8 | Artisan and Service Castes | — | — | — | — | 3 | 3 |
| | | | | | | 100.0 | 100.0 |
| 9 | Harijans | — | — | — | — | 4 | 4 |
| | | | | | | 100.0 | 100.0 |
| 10 | Mappilas | — | 1 | — | — | 1 | 2 |
| | | | 50.0 | | | 50.0 | 100.0 |
| | Total | 33 | 14 | 18 | 10 | 128 | 203 |
| | | 16.3 | 6.9 | 8.9 | 4.9 | 63.1 | 100.0 |

**Note:** Dash indicates nil.
**Source:** Tabulated from the field-data.

Thus, the loss of wet land has been of greater magnitude. It is not only the proportion of loser-households that is higher, but also the proportion of households which have lost a higher proportion of their holdings.

Among the remaining groups, it is only to the Nayars that the loss has been of some significance, that too only of wet land, though compared to the great loss by the Nambutiris it is far less significant. Of the 23 pre-reforms wet land owning households among the Nayars 10 have lost land, with 7 of them losing between 61 per cent and 100 per cent of their holdings.

Tables 6.5 and 6.6 show the wet and garden lands gained by the different groups through the Tribunals, expressed as a

**Table 6.4** ***Households by Loss of Garden Land Through Tribunals (Households in Number and Percentage)***

| *Sl. No.* | *Caste/Community* | *Percentage Loss of Land* | | | | | *Total* |
|---|---|---|---|---|---|---|---|
| | | *81 to 100* | *61 to 80* | *41 to 60* | *Up to 40* | *Nil* | |
| 1 | Nambutiris | 4 | 6 | 12 | 27 | 23 | 72 |
| | | 5.6 | 8.3 | 16.7 | 37.5 | 31.9 | 100.0 |
| 2 | Ambalavasis | — | 1 | 1 | — | 27 | 29 |
| | | | 3.4 | 3.4 | | 93.1 | 100.0 |
| 3 | Nayars | 1 | — | 1 | 1 | 37 | 40 |
| | | 2.5 | | 2.5 | 2.5 | 92.5 | 100.0 |
| 4 | Maniyanis | — | — | — | — | 103 | 103 |
| | | | | | | 100.0 | 100.0 |
| 5 | Muvaris | — | — | — | — | 24 | 24 |
| | | | | | | 100.0 | 100.0 |
| 6 | Vaniyas | — | — | — | 1 | 52 | 53 |
| | | | | | 1.8 | 98.1 | 100.0 |
| 7 | Tiyyas | — | — | — | 5 | 133 | 138 |
| | | | | | 3.6 | 96.4 | 100.0 |
| 8 | Artisan and Service Castes | — | — | — | — | 35 | 35 |
| | | | | | | 100.0 | 100.0 |
| 9 | Harijans | — | — | — | — | 35 | 35 |
| | | | | | | 100.0 | 100.0 |
| 10 | Mappilas | — | — | — | — | 23 | 23 |
| | | | | | | 100.0 | 100.0 |
| | Total | 5 | 7 | 14 | 34 | 492 | 552 |
| | | 0.9 | 1.3 | 2.5 | 6.2 | 89.1 | 100.0 |

**Note:** Dash indicates nil.
**Source:** Tabulated from the field-data.

percentage of the total wet and garden lands owned by them in the post-reforms period. As may be seen from the Tables, the gain has been great for as many as five groups, namely, the Ambalavasis, Nayars, Maniyanis, Vaniyas, and the Tiyyas. Among them, between 59 per cent and 79 per cent of the wet land owning households, and 28 per cent and 45 per cent of the garden land owning households of the post-reforms period have gained land, with between 85 per cent and 96 per cent of the former, and 64 per cent and 82 per cent of the latter gaining between 41 per cent and 100 per cent of their holdings. Of the households which have gained wet land, between

**Table 6.5** ***Households by Gain of Wet Land Through Tribunals (Households in Number and Percentage)***

| *Sl. No.* | *Caste/Community* | *Percentage Gain of Land* | | | | | *Total* |
|---|---|---|---|---|---|---|---|
| | | *81 to 100* | *61 to 80* | *41 to 60* | *Up to 40* | *Nil* | |
| 1 | Nambutiris | — | — | — | — | 37 | 37 |
| | | | | | | 100.0 | 100.0 |
| 2 | Ambalavasis | 10 | 1 | — | 1 | 7 | 19 |
| | | 52.6 | 5.3 | | 5.3 | 36.8 | 100.0 |
| 3 | Nayars | 14 | 2 | 1 | 3 | 13 | 33 |
| | | 42.4 | 6.1 | 3.0 | 9.1 | 39.4 | 100.0 |
| 4 | Maniyanis | 36 | 4 | 6 | 2 | 34 | 82 |
| | | 43.9 | 4.9 | 7.3 | 2.4 | 41.5 | 100.0 |
| 5 | Muvaris | 3 | 1 | — | — | 1 | 5 |
| | | 60.0 | 20.0 | | | 20.0 | 100.0 |
| 6 | Vaniyas | 23 | 5 | 4 | 2 | 9 | 43 |
| | | 53.5 | 11.6 | 9.3 | 4.7 | 20.9 | 100.0 |
| 7 | Tiyyas | 73 | 13 | 7 | 6 | 36 | 135 |
| | | 54.1 | 9.6 | 5.2 | 4.4 | 26.7 | 100.0 |
| 8 | Artisan and Service Castes | 5 | 1 | — | — | 4 | 10 |
| | | 50.0 | 10.0 | | | 40.0 | 100.0 |
| 9 | Harijans | 1 | — | — | — | 4 | 5 |
| | | 20.0 | | | | 80.0 | 100.0 |
| 10 | Mappilas | 1 | 1 | — | — | 2 | 4 |
| | | 25.0 | 25.0 | | | 50.0 | 100.0 |
| | Total | 166 | 28 | 18 | 14 | 147 | 373 |
| | | 44.5 | 7.5 | 4.8 | 3.8 | 39.4 | 100.0 |

**Note:** Dash indicates nil.
**Source:** Tabulated from the field-data.

**Table 6.6** ***Households by Gain of Garden Land Through Tribunals (Households in Number and Percentage)***

| *Sl. No.* | *Caste/Community* | *Percentage Gain of Land* | | | | | *Total* |
|---|---|---|---|---|---|---|---|
| | | *81 to 100* | *61 to 80* | *41 to 60* | *Up to 40* | *Nil* | |
| 1 | Nambutiris | 1 | — | — | 1 | 73 | 75 |
| | | 1.3 | | | 1.3 | 97.3 | 100.0 |
| 2 | Ambalavasis | 6 | 1 | — | 2 | 23 | 32 |
| | | 18.8 | 3.1 | | 6.3 | 71.8 | 100.0 |
| 3 | Nayars | 12 | 2 | 1 | 7 | 36 | 58 |
| | | 20.7 | 3.4 | 1.7 | 12.1 | 62.1 | 100.0 |
| 4 | Maniyanis | 24 | 2 | 3 | 16 | 105 | 150 |
| | | 16.0 | 1.3 | 2.0 | 10.7 | 70.0 | 100.0 |
| 5 | Muvaris | 4 | 1 | 1 | 3 | 27 | 36 |
| | | 11.1 | 2.8 | 2.8 | 8.3 | 75.0 | 100.0 |
| 6 | Vaniyas | 10 | 1 | 7 | 4 | 53 | 75 |
| | | 13.3 | 1.3 | 9.3 | 5.3 | 70.7 | 100.0 |
| 7 | Tiyyas | 60 | 7 | 11 | 28 | 129 | 235 |
| | | 25.5 | 3.0 | 4.7 | 11.9 | 54.9 | 100.0 |
| 8 | Artisan and | 14 | 1 | 3 | 6 | 42 | 66 |
| | Service Castes | 21.2 | 1.5 | 4.5 | 9.1 | 63.6 | 100.0 |
| 9 | Harijans | 6 | — | — | 3 | 38 | 47 |
| | | 12.8 | | | 6.4 | 80.8 | 100.0 |
| 10 | Mappilas | — | — | 1 | 2 | 23 | 26 |
| | | | | 3.8 | 7.7 | 88.5 | 100.0 |
| | Total | 137 | 15 | 27 | 72 | 549 | 800 |
| | | 17.1 | 1.9 | 3.4 | 9.0 | 68.6 | 100.0 |

**Note:** Dash indicates nil.
**Source:** Tabulated from the field-data.

80 per cent and 92 per cent have gained between 61 per cent and 100 per cent of their holdings, whereas of the households which have gained garden land only between 50 per cent and 78 per cent have gained such a high proportion. Thus, corresponding to the loss by the Nambutiris, the gain of wet land by these five groups has been of greater magnitude. It is not only the proportion of gainer-households that is higher, but also the proportion of households which have gained a higher proportion of their holdings.

The gain of wet land has not been significant to the remaining four groups as only between 11 per cent and 15 per cent of their households own any wet land even in the post-reforms period. But, the gain of garden land has been of some significance to the Muvaris, and the artisan and service castes. Of the post-reforms garden land owning households among them, one-fourth (25 per cent) of the former and more than one-third (36 per cent) of the latter have gained garden land through the Tribunals. The relatively very small gain of lands by these four groups has to be explained by the fact that the Act itself was mainly intended to benefit the tenantry and these four groups were not part of the tenantry.

### *Land Tribunals vs. Other Sources*

Besides Land Tribunals there have been other sources through which the different groups could acquire wet and garden lands, namely, purchase and inheritance, and also government assignment in the case of garden land. For bringing out the relative magnitude of land transfers through the Tribunals more clearly, it is necessary to compare these four sources. These sources can be compared in terms of the proportion of households within each group which have acquired wet and garden lands through them, and the proportion of wet and garden lands so acquired. The comparison will be confined to those groups whose gain of wet and garden lands through the Tribunals has been really significant.

Of the post-reforms wet land owning households, those which have purchased any wet land are 3 per cent among the Nayars, and between 20 per cent and 37 per cent among the Ambalavasis, Maniyanis, Vaniyas, and the Tiyyas (Table

6.7); those which have inherited any wet land are between 33 per cent and 48 per cent among all the five groups (Table 6.8); whereas, as already seen, those which have gained wet land through the Tribunals (that too, virtually all their holdings) are between 59 per cent and 79 per cent. Obviously, in the case of all these groups, the magnitude of the gain of wet land through the Tribunals has been far greater than that of acquisition through purchase or inheritance.

As against this, of the post-reforms garden land owning households of these five groups (as also of the Muvaris and the artisan and service castes), those which have acquired any garden land through purchase are between 14 per cent

**Table 6.7** ***Households by Purchase of Wet Land (Households in Number and Percentage)***

| *Sl. No.* | *Caste/Community* | *Percentage of Purchase* | | | | | *Total* |
|---|---|---|---|---|---|---|---|
| | | *81 to 100* | *61 to 80* | *41 to 60* | *Up to 40* | *Nil* | |
| 1 | Nambutiris | 4 | — | 2 | 2 | 29 | 37 |
| | | 10.8 | | 5.4 | 5.4 | 78.4 | 100.0 |
| 2 | Ambalavasis | 3 | — | 2 | 2 | 12 | 19 |
| | | 15.8 | | 10.5 | 10.5 | 63.2 | 100.0 |
| 3 | Nayars | 1 | — | — | — | 32 | 33 |
| | | 3.0 | | | | 97.0 | 100.0 |
| 4 | Maniyanis | 9 | 1 | 1 | 5 | 66 | 82 |
| | | 11.0 | 1.2 | 1.2 | 6.1 | 80.5 | 100.0 |
| 5 | Muvaris | 1 | — | — | — | 4 | 5 |
| | | 20.0 | | | | 80.0 | 100.0 |
| 6 | Vaniyas | 3 | — | 2 | 5 | 33 | 43 |
| | | 7.0 | | 4.7 | 11.6 | 76.7 | 100.0 |
| 7 | Tiyyas | 7 | 3 | 2 | 17 | 106 | 135 |
| | | 5.2 | 2.2 | 1.5 | 12.6 | 78.5 | 100.0 |
| 8 | Artisan and | 2 | — | — | — | 8 | 10 |
| | Service Castes | 20.0 | | | | 80.0 | 100.0 |
| 9 | Harijans | — | — | — | — | 5 | 5 |
| | | | | | | 100.0 | 100.0 |
| 10 | Mappilas | 1 | — | — | 1 | 2 | 4 |
| | | 25.0 | | | 25.0 | 50.0 | 100.0 |
| | Total | 31 | 4 | 9 | 32 | 297 | 373 |
| | | 8.3 | 1.1 | 2.4 | 8.6 | 79.6 | 100.0 |

**Note:** Dash indicates nil.
**Source:** Tabulated from the field-data.

**Table 6.8 *Households by Inheritance of Wet Land (Households in Number and Percentage)***

| *Sl. No.* | *Caste/Community* | *Percentage of Inheritance* | | | | | *Total* |
|---|---|---|---|---|---|---|---|
| | | *81 to 100* | *61 to 80* | *41 to 60* | *Up to 40* | *Nil* | |
| 1 | Nambutiris | 31 | — | 2 | 1 | 3 | 37 |
| | | 83.8 | | 5.4 | 2.7 | 8.1 | 100.0 |
| 2 | Ambalavasis | 2 | 1 | 1 | 3 | 12 | 19 |
| | | 10.5 | 5.3 | 5.3 | 15.8 | 63.2 | 100.0 |
| 3 | Nayars | 12 | 2 | 1 | 1 | 17 | 33 |
| | | 36.4 | 6.1 | 3.0 | 3.0 | 51.5 | 100.0 |
| 4 | Maniyanis | 15 | 2 | 3 | 8 | 54 | 82 |
| | | 18.3 | 2.4 | 3.7 | 9.8 | 65.9 | 100.0 |
| 5 | Muvaris | — | — | — | 1 | 4 | 5 |
| | | | | | 20.0 | 80.0 | 100.0 |
| 6 | Vaniyas | 5 | 2 | 2 | 5 | 29 | 43 |
| | | 11.6 | 4.7 | 4.7 | 11.6 | 67.4 | 100.0 |
| 7 | Tiyyas | 28 | 3 | 4 | 16 | 84 | 135 |
| | | 20.7 | 2.2 | 3.0 | 11.9 | 62.2 | 100.0 |
| 8 | Artisan and Service Castes | 2 | — | 1 | — | 7 | 10 |
| | | 20.0 | | 10.0 | | 70.0 | 100.0 |
| 9 | Harijans | 4 | — | — | — | 1 | 5 |
| | | 80.0 | | | | 20.0 | 100.0 |
| 10 | Mappilas | 1 | — | — | — | 3 | 4 |
| | | 25.0 | | | | 75.0 | 100.0 |
| | Total | 100 | 10 | 14 | 35 | 214 | 373 |
| | | 26.8 | 2.7 | 3.7 | 9.4 | 57.4 | 100.0 |

**Note:** Dash indicates nil.
**Source:** Tabulated from the field-data.

and 40 per cent (Table 6.9), and those which have acquired any garden land through government assignment are between 11 per cent and 36 per cent (Table 6.10), whereas those which have acquired any garden land through inheritance are between 35 per cent and 62 per cent (Table 6.11) and, as already seen, those which have gained garden land through the Tribunals are between 25 per cent and 45 per cent. Thus, the magnitude of the acquisition of garden land by these five groups has been greater through inheritance than that through any other source, including the Tribunals.

Tables 6.12 and 6.13 show the wet and garden lands acquired

**Table 6.9** ***Households by Purchase of Garden Land (Households in Number and Percentage)***

| *Sl. No.* | *Caste/Community* | *Percentage of Purchase* | | | | | *Total* |
|---|---|---|---|---|---|---|---|
| | | *81 to 100* | *61 to 80* | *41 to 60* | *Up to 40* | *Nil* | |
| 1 | Nambutiris | 5 | 1 | — | 5 | 64 | 75 |
| | | 6.7 | 1.3 | | 6.7 | 85.3 | 100.0 |
| 2 | Ambalavasis | 6 | — | 1 | — | 25 | 32 |
| | | 18.8 | | 3.1 | | 78.1 | 100.0 |
| 3 | Nayars | 4 | 1 | 1 | 3 | 49 | 58 |
| | | 6.9 | 1.7 | 1.7 | 5.2 | 84.5 | 100.0 |
| 4 | Maniyanis | 15 | 6 | 3 | 14 | 112 | 150 |
| | | 10.0 | 4.0 | 2.0 | 9.3 | 74.7 | 100.0 |
| 5 | Muvaris | 3 | — | 2 | — | 31 | 36 |
| | | 8.3 | | 5.6 | | 86.1 | 100.0 |
| 6 | Vaniyas | 10 | 4 | 10 | 6 | 45 | 75 |
| | | 13.3 | 5.3 | 13.3 | 8.0 | 60.0 | 100.0 |
| 7 | Tiyyas | 19 | 3 | 2 | 13 | 198 | 235 |
| | | 8.1 | 1.3 | 0.9 | 5.5 | 84.3 | 100.0 |
| 8 | Artisan and | 14 | 2 | 2 | 2 | 46 | 66 |
| | Service Castes | 21.2 | 3.0 | 3.0 | 3.0 | 69.7 | 100.0 |
| 9 | Harijans | 2 | — | — | 5 | 40 | 47 |
| | | 4.3 | | | 10.6 | 85.1 | 100.0 |
| 10 | Mappilas | 7 | 2 | — | 3 | 14 | 26 |
| | | 26.9 | 7.7 | | 11.5 | 53.8 | 100.0 |
| | Total | 85 | 19 | 21 | 51 | 624 | 800 |
| | | 10.6 | 2.4 | 2.6 | 6.4 | 78.0 | 100.0 |

**Note:** Dash indicates nil.
**Source:** Tabulated from the field-data.

by the different groups through the different sources, expressed as percentage of the total wet and garden lands owned by them in the post-reforms period.

Of the total wet land owned by the Ambalavasis, Nayars, Maniyanis, Vaniyas, and the Tiyyas, between 51 per cent and 74 per cent has been gained through the Tribunals, whereas only between 18 per cent and 48 per cent has been acquired through inheritance, and only between 1 per cent and 15 per cent has been acquired through purchase, or only between 26 per cent and 49 per cent has been acquired through inheritance and purchase combined.

**Table 6.10** ***Households by Government Assignment of Garden Land (Households in Number and Percentage)***

| *Sl. No.* | *Caste/Community* | *Percentage of Assignment* | | | | | *Total* |
|---|---|---|---|---|---|---|---|
| | | *81 to 100* | *61 to 80* | *41 to 60* | *Up to 40* | *Nil* | |
| 1 | Nambutiris | — | 1 | — | 1 | 73 | 75 |
| | | | 1.3 | | 1.3 | 97.3 | 100.0 |
| 2 | Ambalavasis | 4 | 2 | — | 2 | 24 | 32 |
| | | 12.5 | 6.3 | | 6.3 | 75.0 | 100.0 |
| 3 | Nayars | 4 | 2 | — | 3 | 49 | 58 |
| | | 6.9 | 3.5 | | 5.2 | 84.5 | 100.0 |
| 4 | Maniyanis | 19 | 3 | 3 | 7 | 118 | 150 |
| | | 12.7 | 2.0 | 2.0 | 4.7 | 78.7 | 100.0 |
| 5 | Muvaris | 10 | 1 | 2 | — | 23 | 36 |
| | | 27.8 | 2.8 | 5.5 | | 63.9 | 100.0 |
| 6 | Vaniyas | 2 | 4 | 2 | — | 67 | 75 |
| | | 2.7 | 5.3 | 2.7 | | 89.3 | 100.0 |
| 7 | Tiyyas | 20 | 7 | 10 | 8 | 190 | 235 |
| | | 8.5 | 3.0 | 4.3 | 3.4 | 80.8 | 100.0 |
| 8 | Artisan and | 9 | 3 | 2 | 1 | 51 | 66 |
| | Service Castes | 13.6 | 4.6 | 3.0 | 1.5 | 77.3 | 100.0 |
| 9 | Harijans | 11 | 1 | 2 | 3 | 30 | 47 |
| | | 23.4 | 2.1 | 4.3 | 6.4 | 63.8 | 100.0 |
| 10 | Mappilas | 2 | — | — | 3 | 21 | 26 |
| | | 7.7 | | | 11.5 | 80.8 | 100.0 |
| | Total | 81 | 24 | 21 | 28 | 646 | 800 |
| | | 10.1 | 3.0 | 2.6 | 3.5 | 80.8 | 100.0 |

**Note:** Dash indicates nil.
**Source:** Tabulated from the field-data.

As against this, of the total garden land owned by these groups (as also the Muvaris and the artisan and service castes), between 24 per cent and 68 per cent has been acquired through inheritance, between 5 per cent and 29 per cent has been acquired through purchase, between 9 per cent and 34 per cent has been acquired through government assignment, and only between 14 per cent and 23 per cent has been gained through the Tribunals.

The relatively low proportion of wet land acquired through sources other than Land Tribunals by the Ambalavasis, Nayars, Maniyanis, Vaniyas and the Tiyyas indicates not

**Table 6.11** ***Households by Inheritance of Garden Land (Households in Number and Percentage)***

| *Sl. No.* | *Caste/Community* | *Percentage of Inheritance* | | | | | *Total* |
|---|---|---|---|---|---|---|---|
| | | *81 to 100* | *61 to 80* | *41 to 60* | *Up to 40* | *Nil* | |
| 1 | Nambutiris | 62 | 4 | — | 2 | 7 | 75 |
| | | 82.7 | 5.3 | | 2.7 | 9.3 | 100.0 |
| 2 | Ambalavasis | 11 | 1 | 3 | 3 | 14 | 32 |
| | | 34.4 | 3.1 | 9.4 | 9.4 | 43.7 | 100.0 |
| 3 | Nayars | 24 | 6 | 2 | 4 | 22 | 58 |
| | | 41.4 | 10.3 | 3.5 | 6.9 | 37.9 | 100.0 |
| 4 | Maniyanis | 53 | 11 | 6 | 11 | 69 | 150 |
| | | 35.3 | 7.3 | 4.0 | 7.3 | 46.0 | 100.0 |
| 5 | Muvaris | 12 | 2 | 1 | 2 | 19 | 36 |
| | | 33.3 | 5.6 | 2.8 | 5.6 | 52.8 | 100.0 |
| 6 | Vaniyas | 27 | 3 | 5 | 11 | 29 | 75 |
| | | 36.0 | 4.0 | 6.7 | 14.7 | 38.7 | 100.0 |
| 7 | Tiyyas | 71 | 14 | 13 | 16 | 121 | 235 |
| | | 30.2 | 6.0 | 5.5 | 6.8 | 51.5 | 100.0 |
| 8 | Artisan and | 15 | 1 | 3 | 4 | 43 | 66 |
| | Service Castes | 22.7 | 1.5 | 4.6 | 6.1 | 65.2 | 100.0 |
| 9 | Harijans | 17 | 4 | 2 | 1 | 23 | 47 |
| | | 36.2 | 8.5 | 4.3 | 2.1 | 48.9 | 100.0 |
| 10 | Mappilas | 12 | 2 | — | — | 12 | 26 |
| | | 46.2 | 7.7 | | | 46.2 | 100.0 |
| | Total | 304 | 48 | 35 | 54 | 359 | 800 |
| | | 38.0 | 6.0 | 4.4 | 6.7 | 44.9 | 100.0 |

**Note:** Dash indicates nil.
**Source:** Tabulated from the field-data.

only the greater magnitude of the gain of wet land through the Tribunals, but also the fact that either the *janmis* were jealously guarding their ownership rights in wet land as it was observed earlier, or the cultivating tenants could not afford to buy any wet land, or they were reluctant to do so anticipating the good days of the implementation of the land reforms Act.

As against this, the relatively very high proportion of garden land acquired through sources other than Tribunals points to the important fact that not all the beneficiary households were necessarily landless in the pre-reforms period, and that to those beneficiaries who were owning garden land,

**Table 6.12** ***Ownership of Wet Land by Source of Acquisition (Area in Acres and Percentage)***

| *Sl. No.* | *Caste/Community* | *Land Tribunals* | *Inheritance* | *Purchase* | *Total* |
|---|---|---|---|---|---|
| 1 | Nambutiris | Nil | 40.3 | 3.6 | 43.9 |
| | | | 91.8 | 8.2 | 100.0 |
| 2 | Ambalavasis | 14.0 | 4.4 | 3.2 | 21.6 |
| | | 64.8 | 20.4 | 14.8 | 100.0 |
| 3 | Nayars | 17.0 | 16.0 | 0.4 | 33.4 |
| | | 50.9 | 47.9 | 1.2 | 100.0 |
| 4 | Maniyanis | 39.4 | 18.3 | 5.1 | 62.8 |
| | | 62.7 | 29.1 | 8.1 | 100.0 |
| 5 | Muvaris | 1.8 | 0.2 | 0.1 | 2.1 |
| | | 85.7 | 9.5 | 4.8 | 100.0 |
| 6 | Vaniyas | 23.1 | 5.7 | 2.5 | 31.3 |
| | | 73.8 | 18.2 | 8.0 | 100.0 |
| 7 | Tiyyas | 62.6 | 44.6 | 11.2 | 118.4 |
| | | 52.9 | 37.7 | 9.4 | 100.0 |
| 8 | Artisan and | 2.4 | 1.1 | 1.0 | 4.5 |
| | Service Castes | 53.3 | 24.4 | 22.2 | 100.0 |
| 9 | Harijans | 0.6 | 6.5 | Nil | 7.1 |
| | | 8.5 | 91.5 | | 100.0 |
| 10 | Mappilas | 1.2 | 1.1 | 0.9 | 3.2 |
| | | 37.5 | 34.4 | 28.1 | 100.0 |
| | Total | 162.1 | 138.2 | 28.0 | 328.3 |
| | | 49.4 | 42.1 | 8.5 | 100.0 |

**Source:** Tabulated from the field-data.

Land Tribunals have only been an additional source through which they could augment their holdings. In any case, what is pertinent to note here is that the loss of wet and garden lands by the Nambutiris, and the gain of the same by the Ambalavasis, Nayars, Maniyanis, Vaniyas, and the Tiyyas (to mention only those groups whose loss and gain have been really significant) have brought about drastic changes in the village land system.

## Changes in Land System

These changes can be clearly brought out by a comparison of the land ownership patterns in the pre- and post-reforms

**Table 6.13** ***Ownership of Garden Land by Source of Acquisition (Area in Acres and Percentage)***

| *Sl. No.* | *Caste/Community* | *LT* | *I* | *P* | *GA* | *K* | *S* | *Total* |
|---|---|---|---|---|---|---|---|---|
| 1 | Nambutiris | 3.9 | 292.6 | 26.5 | 7.2 | — | — | 330.2 |
| | | 1.2 | 88.6 | 8.0 | 2.2 | | | 100.0 |
| 2 | Ambalavasis | 12.2 | 35.7 | 10.9 | 19.0 | — | — | 77.8 |
| | | 15.7 | 45.9 | 14.0 | 24.4 | | | 100.0 |
| 3 | Nayars | 31.5 | 115.9 | 8.9 | 14.8 | — | — | 171.1 |
| | | 18.4 | 67.7 | 5.2 | 8.6 | | | 100.0 |
| 4 | Maniyanis | 45.6 | 133.1 | 51.6 | 70.8 | 0.1 | 1.0 | 302.2 |
| | | 15.1 | 44.0 | 17.1 | 23.4 | 0.0 | 0.3 | 100.0 |
| 5 | Muvaris | 6.2 | 17.4 | 6.1 | 15.5 | — | — | 45.2 |
| | | 13.7 | 38.5 | 13.5 | 34.3 | | | 100.0 |
| 6 | Vaniyas | 21.3 | 50.9 | 36.2 | 14.5 | — | — | 122.9 |
| | | 17.3 | 41.4 | 29.5 | 11.8 | | | 100.0 |
| 7 | Tiyyas | 130.8 | 296.0 | 51.9 | 85.6 | 0.1 | 4.2 | 568.6 |
| | | 23.0 | 52.1 | 9.1 | 15.1 | 0.0 | 0.7 | 100.0 |
| 8 | Artisan and | 19.0 | 20.7 | 21.9 | 24.3 | 0.1 | 0.4 | 86.4 |
| | Service Castes | 22.0 | 24.0 | 25.3 | 28.1 | 0.1 | 0.5 | 100.0 |
| 9 | Harijans | 5.6 | 84.2 | 12.3 | 29.0 | 0.3 | 0.9 | 132.3 |
| | | 4.2 | 63.6 | 9.3 | 21.9 | 0.2 | 0.7 | 100.0 |
| 10 | Mappilas | 2.4 | 20.4 | 21.1 | 10.5 | — | — | 54.4 |
| | | 4.4 | 37.5 | 38.8 | 19.3 | | | 100.0 |
| | Total | 278.5 | 1066.9 | 247.4 | 291.2 | 0.6 | 6.5 | 1891.1 |
| | | 14.7 | 56.4 | 13.1 | 15.4 | 0.0 | 0.3 | 100.0 |

**Note:** Dash indicates nil. LT: Land Tribunals; I: inheritance; P: purchase; GA: government assignments; K: *kudikidappu*; S: surplus distributed.

**Source:** Tabulated from the field-data.

periods in terms of the proportions of wet and garden land owning households among the different groups, the mean size of wet and garden land holdings per household within each group, and the proportions of wet and garden lands owned by these groups. These are shown in Tables 6.14, 6.15, and 6.16 respectively. The changes are also brought out very strikingly in the accompanying histograms on the proportion of land owned in the pre- and post-reforms periods by the groups which are important in relation to the land reforms as either losers or gainers of land.

Among the Nambutiris, only about half (49 per cent) the

**Table 6.14 *Households Owning Wet Land in Pre- and Post-Reforms Periods***

| *Sl. No.* | *Caste/Community* | *Pre-reforms* | | *Post-reforms* | | *N* |
|---|---|---|---|---|---|---|
| | | *Households* | *Percentage* | *Households* | *Percentage* | |
| 1 | Nambutiris | 64 | 85.3 | 37 | 49.3 | 75 |
| 2 | Ambalavasis | 9 | 28.1 | 19 | 59.4 | 32 |
| 3 | Nayars | 23 | 39.7 | 33 | 56.9 | 58 |
| 4 | Maniyanis | 32 | 21.3 | 82 | 54.7 | 150 |
| 5 | Muvaris | Nil | Nil | 5 | 13.9 | 36 |
| 6 | Vaniyas | 16 | 21.3 | 43 | 57.3 | 75 |
| 7 | Tiyyas | 50 | 21.3 | 135 | 57.5 | 235 |
| 8 | Artisan and Service Castes | 3 | 4.5 | 10 | 15.2 | 66 |
| 9 | Harijans | 4 | 8.5 | 5 | 10.6 | 47 |
| 10 | Mappilas | 2 | 7.7 | 4 | 15.4 | 26 |
| | Total | 203 | 25.4 | 373 | 46.7 | 800 |

**Source:** Tabulated from the field-data.

**Table 6.15** *Mean Size of Land Holdings in Pre- and Post-Reforms Periods (Land Holdings in Acres)*

| Sl. No. | Caste/Community | Wet Land | | | | Garden Land | | | |
|---|---|---|---|---|---|---|---|---|---|
| | | Pre-reforms | | Post-reforms | | Pre-reforms | | Post-reforms | |
| | | MS | SD | MS | SD | MS | SD | MS | SD |
| 1 | Nambutiris | 3.50 | 9.30 | 0.59 | 0.92 | 7.72 | 13.61 | 4.41 | 4.13 |
| 2 | Ambalavasis | 0.35 | 1.71 | 0.68 | 0.75 | 1.67 | 1.77 | 2.43 | 1.46 |
| 3 | Nayars | 0.82 | 1.67 | 0.58 | 0.82 | 2.44 | 3.77 | 2.95 | 3.59 |
| 4 | Maniyanis | 0.11 | 0.27 | 0.42 | 0.62 | 1.45 | 1.72 | 2.01 | 1.94 |
| 5 | Muvaris | Nil | Nil | 0.06 | 0.17 | 0.81 | 0.78 | 1.26 | 0.70 |
| 6 | Vaniyas | 0.10 | 0.37 | 0.42 | 0.58 | 1.07 | 1.55 | 1.64 | 1.67 |
| 7 | Tiyyas | 0.17 | 0.61 | 0.50 | 0.85 | 1.58 | 4.38 | 2.42 | 5.13 |
| 8 | Artisan and Service Castes | 0.02 | 0.10 | 0.07 | 0.18 | 0.68 | 1.09 | 1.31 | 1.32 |
| 9 | Harijans | 0.14 | 0.57 | 0.15 | 0.56 | 2.37 | 3.65 | 2.82 | 4.27 |
| 10 | Mappilas | 0.06 | 0.23 | 0.13 | 0.35 | 1.58 | 1.40 | 2.10 | 1.80 |

**Source:** Tabulated from the field-data.

**Table 6.16** ***Ownership of Wet and Garden Lands in Pre- and Post-Reforms Periods (Area in Acres and Percentage)***

| *Sl. No.* | *Caste/Community* | *Pre-reforms* | | *Post-reforms* | | *Households in Percentage* |
|---|---|---|---|---|---|---|
| | | *Wet Land* | *Garden Land* | *Wet Land* | *Garden Land* | |
| 1. | Nambutiris | 262.5 | 578.9 | 43.9 | 330.2 | |
| | | 66.4 | 34.7 | 13.4 | 17.5 | 9.4 |
| 2 | Ambalavasis | 11.4 | 53.6 | 21.6 | 77.8 | |
| | | 2.9 | 3.2 | 6.6 | 4.1 | 4.0 |
| 3 | Nayars | 47.6 | 141.3 | 33.4 | 171.1 | |
| | | 12.1 | 8.5 | 10.2 | 9.0 | 7.2 |
| 4 | Maniyanis | 1 5.8 | 216.9 | 62.8 | 302.2 | |
| | | 4.0 | 13.0 | 19.1 | 16.0 | 18.7 |
| 5 | Muvaris | Nil | 29.0 | 2.1 | 45.2 | |
| | | | 1.7 | 0.6 | 2.4 | 4.5 |
| 6 | Vaniyas | 7.8 | 80.3 | 31.3 | 122.9 | |
| | | 2.0 | 4.8 | 9.5 | 6.5 | 9.4 |
| 7 | Tiyyas | 40.5 | 372.3 | 118.4 | 568.6 | |
| | | 10.3 | 22.3 | 36.1 | 30.1 | 29.4 |
| 8 | Artisan and | 1.4 | 45.1 | 4.5 | 86.4 | |
| | Service Castes | 0.3 | 2.7 | 1.4 | 4.6 | 8.2 |
| 9 | Harijans | 6.5 | 111.4 | 7.1 | 132.3 | |
| | | 1.6 | 6.7 | 2.2 | 7.0 | 5.9 |
| 10 | Mappilas | 1.5 | 41.1 | 3.2 | 54.4 | |
| | | 0.4 | 2.5 | 1.0 | 2.9 | 3.2 |
| | Total | 395.0 | 1669.9 | 328.3 | 1891.1 | |
| | | 100.0 | 100.0 | 100.0 | 100.0 | 100.0 |

**Source:** Tabulated from the field-data.

households own any wet land now, with only 0.6 acre as the mean size. Earlier more than four-fifths (85 per cent) had such ownership, with 3.5 acres as the mean size. Though all the Nambutiri households own garden land now, the mean size is only 4.4 acres as against the earlier 7.7 acres. Their share in the total land ownership now is only 13.4 per cent of the wet land and 17.5 per cent of the garden land,as against the earlier 66 per cent and 35 per cent respectively.

Among the Ambalavasis, Nayars, Maniyanis, Vaniyas, and the Tiyyas, between 55 per cent and 59 per cent of the households own wet land now, and the mean size is between 0.4 and 0.8 acre. Earlier only 40 per cent of the households of

**Figure 1** ***Caste Size and Wet Land Ownership Pre- and Post-Reforms Periods***

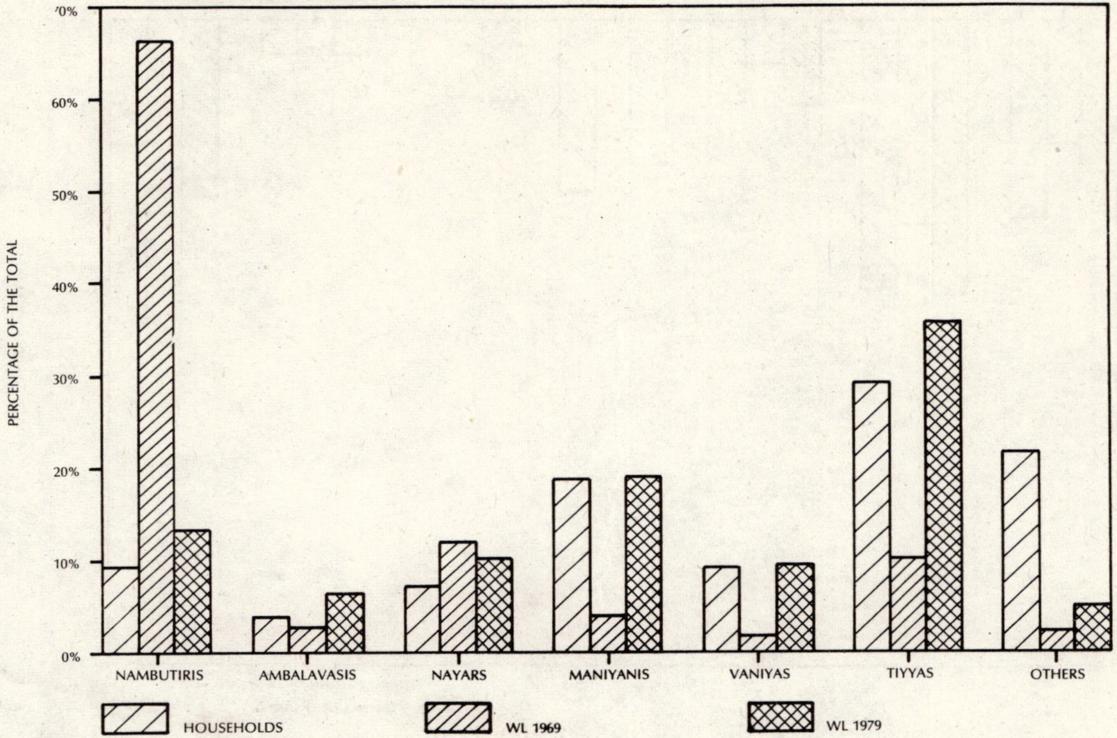

**Figure: 2** ***Caste Size and Gardenland Ownership Pre- and Post-Reforms Periods***

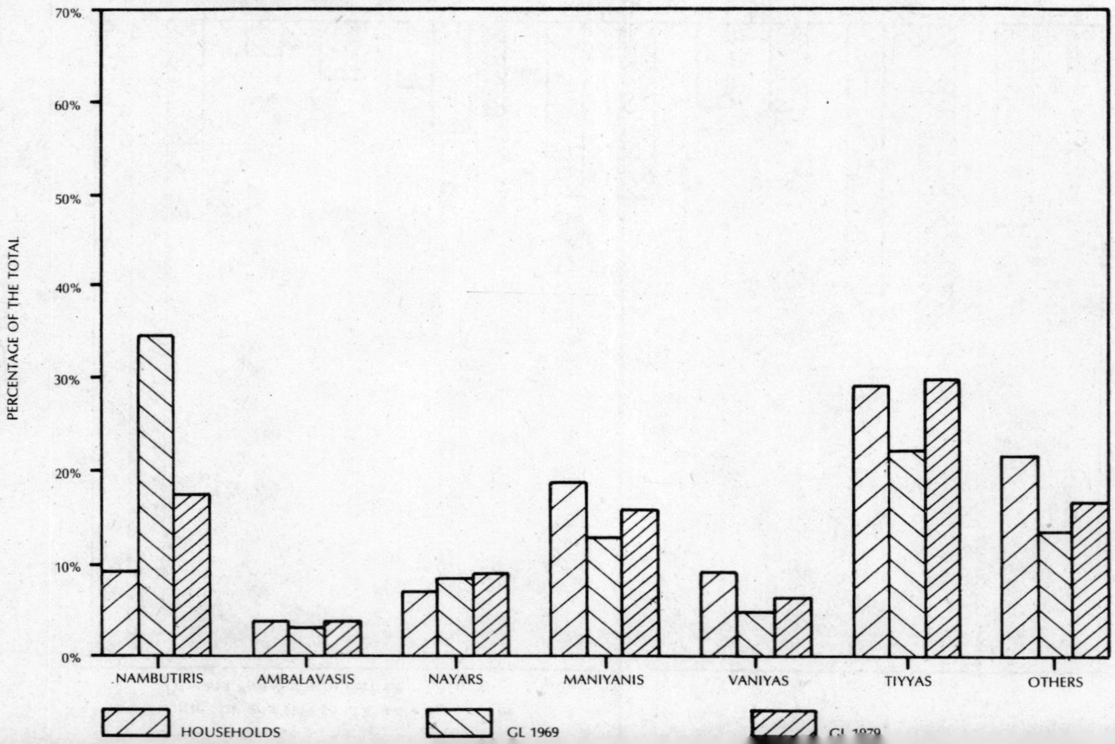

the Nayars, and 21 per cent each among the other four groups had such ownership. The mean size was 0.8 acre among the Nayars, 0.4 acre among the Ambalavasis, and much less among the others. All the households of these five groups (as also of all the remaining groups) now own garden land, and the mean size is between 1.6 and 3.0 acres. Earlier only between 53 per cent and 91 per cent of the households had such ownership, and the mean size was only between 0.7 acre and 2.4 acres.

The land system now is thus strikingly different from what it was in the pre-reforms period. The change is characterised by status-loss by the Nambutiris, status retention by the Nayars, and status-gain by the Ambalavasis, Maniyanis, Vaniyas, and the Tiyyas. Status loss by the Nambutiris is the consequence of the steep fall in their wet land owning households, in the mean size of wet and garden lands owned, and in the proportions of wet and garden lands owned. Status retention by the Nayars is the result of the rise in their wet and garden land owning households, in the garden land owned, and in its mean size, and fall in the mean size of wet land, and in the proportion of wet land owned. The gain in status by the Ambalavasis, Maniyanis, Vaniyas, and the Tiyyas, is because of the sharp rise in their wet and garden land owning households, mean size, and in the overall size of wet and garden lands owned. The proportion of wet and garden lands owned by these six groups in the post-reforms period is almost equal to the proportion of their households. Consequently, the land system is no longer characterised by the extreme concentration of lands in a single group, and there is no longer any caste-class correspondence in the land ownership in the village.

## APPENDIX

### Note on Village Survey

As in the rest of Kerala, the village is large and its houses are scattered. This poses serious problems for field-work. Its size, both geographic and demographic, is large and unwieldy and it is not easy to determine where its boundaries begin and end. In fact, there are different sectors within the village (such as, Vellachal, Kizhakke Kara, Padinhare Kara, Puthilottu, Mappidacheri, Pollapoyil, Vengappara, Anikkadi, Kunhippara, Settlement Area, and Kookkanam), each of which is in the nature of a nucleated village. Further, the division of the village as a revenue, administrative, and census unit is not coterminous with the village as a social unit. The latter often transcends the village boundaries.

The investigation conducted in the village was in the first half of 1979, in two field-visits each of three months duration. It was in the nature of in-depth interviews with knowledgeable persons in and around the village, observations of the village social life, and the collection of quantitative data with the help of a schedule. These sought detailed information on the history of the third phase of the peasant struggles in different parts of Malabar including this village as recounted by the participants themselves; on the land and society in the village in the past; on the implementation of the 1969 Act and its effects on both; and on other related issues.

While the collection of qualitative information through interviews and observation did not pose any serious problem, that of quantitative data through schedules turned out to be difficult. Although rapport was established with important village leaders well before the first field-trip, many of the villagers were initially indifferent or gave only superficial answers. There were several reasons for this. The villagers in general were resentful of being interrogated repeatedly and within a short span of time for various purposes—first to study the Teyyan-cult, by a group of American scholars who had camped at the village in 1976–77; then to study the caste practices among the Tiyyas, by two Dutch scholars; and for data collection for the revenue and statistics departments of the Kerala government, by some investigators; and finally for this study. Apart from this, some of the villagers expected to be paid for parting with any information, following a precedent set by the American scholars who had made payments to the Teyyam-dancers for demonstration performances. Some resented questions on caste, politics, and party affiliations, and yet a few withheld information protesting against the panchayat office for not complying with their request for digging wells. Among the various groups, the Nambutiris were initially the least responsive. For fear of the communists, the communist government, the tax collectors, and the 'removal of untouchability squad,' many of them first refused to even come out of their houses.

While these initial hurdles were gradually surmounted by repeated contacts, persuasion, and the goodwill of some of the villagers themselves, there were still more problems. Initially, the idea was to collect quantitative data from all households. The difficulties of such a venture were realised soon enough from the vast area of the village and the large number of its scattered households to be reached on foot, and from the necessity of winding up work early in the day because of the lack of *pucca* roads, electrification and transport facilities.*

* At the time of field investigation the village did not have any pucca roads, while its electrification was about to be begun. Though two buses were plying through it connecting it with the nearest town of Payyanur, they

But, its impossibility could be realised only towards the end. In many Mappila households men were mostly away and women were either too reluctant to talk to a stranger or not able to supply the relevant information. In several of the households of the Tiyyas, Vaniyas, and the Harijans, the heads were away in the forenoons, while in the afternoons they were found to be under the influence of liquor, thanks to the nearly half a dozen toddy-shops in different parts of the village, mostly in the areas where these groups are concentrated. In some other cases the heads worked in distant places and returned home only late in the evening. Contacting them at the work place did not yield any result as they were either unwilling to spare time answering questions, or unable to remember all the required details. In a few cases the heads worked outside the village and returned only periodically. The other family members were not in a position to supply the various details, especially on land ownership and land transfers. Considering the limited time and the distance to be covered on foot to reach the non-available households scattered throughout the village, it was not possible to repeat the house-calls more than a few times. As a result, the percentage of no-response households turned out to be quite high.

However, as the house-calls made were with the help of the voters' list supplied by the panchayat office and in the order in which the houses were listed (namely, serially, within each ward), and as the houses within each ward were generally segregated caste-wise, the high percentage of no-response cases does not underrepresent any group, especially those significant in relation to land. This will be evident from the distribution of respondent and non-respondent households in the village as shown in Table 6.17.

---

were restricted to the northern side of the village, and to one trip in the morning and one in the evening. The villagers generally covered about 5 Kms on foot to reach the nearest bus stop at Karivellur.

**Table 6.17** ***Respondent and Non-Respondent Households by Caste (Households in Number and Percentage)***

| *Sl. No.* | *Caste/Community* | *Households* | | *Total* |
|---|---|---|---|---|
| | | *Respondent* | *Non-respondent* | |
| 1 | Nambutiris | 75 | Nil | 75 |
| | | 100.0 | | 100.0 |
| 2 | Ambalavasis | 32 | 11 | 43 |
| | | 74.4 | 25.6 | 100.0 |
| 3 | Nayars | 58 | 23 | 81 |
| | | 71.6 | 28.4 | 100.0 |
| 4 | Maniyanis | 150 | 50 | 200 |
| | | 75.0 | 25.0 | 100.0 |
| 5 | Muvaris | 36 | 13 | 49 |
| | | 73.5 | 26.5 | 100.0 |
| 6 | Vaniyas | 75 | 40 | 115 |
| | | 65.2 | 34.8 | 100.0 |
| 7 | Tiyyas | 235 | 165 | 400 |
| | | 58.8 | 41.2 | 100.0 |
| 8 | Artisan and | 66 | 9 | 75 |
| | Service Castes | 88.0 | 12.0 | 100.0 |
| 9 | Harijans | 47 | 25 | 72 |
| | | 65.3 | 34.7 | 100.0 |
| 10 | Mappilas | 26 | 26 | 52 |
| | | 50.0 | 50.0 | 100.0 |
| | Total | 800 | 362 | 1,162 |
| | | 68.9 | 31.1 | 100.0 |

**Source:** Tabulated from the field-data.

# 7

# Land Reforms and Social Change, 1970–79

The changes in the land system in Kerala consequent on the implementation of the KLRAA of 1969 are expected to have created conditions for other perceptible changes in those aspects of rural society which were directly and immediately related to the land system. An overall assessment of these changes would involve understanding their different micro-versions, keeping in view the variations in the socio-economic conditions in different parts of the state. This chapter provides one such version in the context of Kodakkad against the backdrop of the changes in its land system.

The chapter is concerned with those social groups of the village in relation to which significant changes have been perceived in the land system, and with those aspects of village society in which these changes have been found to be significant. The groups concerned are the Nambutiris, Ambalavasis, Nayars, Maniyanis, Vaniyas, and the Tiyyas whose households together account for 78 per cent of the households of all the ten groups. In the pre-reforms period, the wet and garden lands owned by them accounted for about 98 per cent and 86 per cent respectively of those owned by the ten groups together. In the post-reforms period these account for about 95 per cent and 83 per cent respectively. The aspects concerned are the economic structure, occupational structure, household composition, religious practices, power structure, and social relations. The chapter will focus on the Nambutiris and the Tiyyas, the groups in relation to which the changes have been found to be the most significant, the former as the only major losers and the latter as the only major gainers among the former polluting castes.

In understanding this focus it is at the outset necessary to bear in mind that the beginnings of change in relation to these two groups well ante-date the introduction of the recent land reforms and were part of the general changes taking place in the larger society. A discussion of these changes is beyond the scope of this study. Suffice to say that starting from the early part of this century, the social organisation of the Nambutiris had begun to face internal challenges through the *Nambutiri Yogakshema Mahasabha* (the Nambutiri Welfare Association), and the *Unni Nambutiri Samajam* (the Nambutiri Youth Movement). These associations had advocated major changes in the Nambutiri community, which included freedom of the Nambutiri boys and girls to take up English education, the right of all Nambutiri males to formal marriage, right to partition of *illam* property, and so on, most of which had come about by the second quarter of this century itself. However, the pace of the earlier changes had been uneven. While some areas experienced rapid change, some others remained backward.[1] As Kodakkad belongs to a region which is strongly traditional, the role of land reforms as a change-agent in it assumes added importance.

Starting from the same period, the Tiyyas had initiated militant agitations against their social disabilities as a lower caste. The changes were uneven in their case also. They were rapid and widespread in south Kerala and slow and uneven in the north. It was also in south Kerala that the *Sree Narayana Dharma Paripalana Yogam* (SNDP Yogam)—the socio-cultural and political organisation of the Tiyyas—emerged as a powerful social force. The SNDP Yogam still does not have any hold in many villages in north Malabar and its nearest

[1] For a review of the Nambutiri reforms movement and its consequences, see Mencher, 1966; Namboodiripad, 1976:21–28.

unit to Kodakkad is in Payyanur town.[2] Considering that the social reforms movement of the Tiyyas did not make much headway in Kodakkad and the traditional social organisation of the Tiyyas was not much affected by it, land reforms as change-agent assumes added importance in their case as well.

## Economic Structure

In striking contrast to the earlier dominance of the Nambutiris and dependence on them by the other five groups (as also the remaining groups), the present economic status of all the six groups is one of relative equality. One of the important indicators of this is the post-reforms income distribution, as shown in Table 7.1. Among the Nambutiris the distribution of households in the three annual income categories of low (up to Rs. 3,000), medium (Rs. 3,000 to Rs. 6,000) and high (Rs. 6,000 and above) is 60 per cent, 20 per cent, and 20 per cent respectively. The corresponding percentages for the Ambalavasis and the Nayars are 63, 25, and 13 and 57, 29 and 14 respectively, while those for the Maniyanis, Vaniyas, and the Tiyyas are 83, 10 and 7, 80, 16, and 4, and 73, 19, and 8 respectively.[3] A comparison of these percentages reveals

[2] For a review of the social reforms movement among the Tiyyas, see Rao, 1979.

[3] The annual income includes income from both land holdings and occupations. Since it is mainly based on figures furnished by the respondents and not on any rigorous computation, the Table gives only an estimate of the income distribution. Village level data on pre-reforms income distribution are not available. As the respondents had not kept any record of their income and as information furnished by them from their memory would not have been reliable, no attempt was made to collect such data. However, since land has been the main resource to all the six groups both in the pre- and post-reforms periods, and the Nambutiris had a virtual monopoly of the village lands in the pre-reforms period, even in the absence of any data it is only too obvious that their income in the pre-reforms period would have been much higher than that of the other five groups. Some indications of this are also available in the 1966–67 LRS data on Cannanore district to which this village belongs. According to the LRS, the distribution of landlord households in the three annual income categories of below Rs. 3,000, Rs. 3,000 to Rs. 6,000 and Rs. 6,000 and above was as 46 per cent, 29 per cent, and 25 per cent respectively. The corresponding percentages for the tenant households were 86, 11, and 2 respectively. See Government of Kerala, 1968:143.

**Table 7.1** ***Households by Annual Income in the Post-Reforms Period (Households in Number and Percentage)***

| *Sl. No.* | *Caste/Community* | *Up to Rs. 3,000* | *Rs. 3,000 to Rs. 6,000* | *Rs. 6,000 and above* | *Total* |
|---|---|---|---|---|---|
| 1 | Nambutiris | 45 | 15 | 15 | 75 |
| | | 60.0 | 20.0 | 20.0 | 100.0 |
| 2 | Ambalavasis | 20 | 8 | 4 | 32 |
| | | 62.5 | 25.0 | 12.5 | 100.0 |
| 3 | Nayars | 33 | 17 | 8 | 58 |
| | | 56.9 | 29.3 | 13.8 | 100.0 |
| 4 | Maniyanis | 125 | 15 | 10 | 150 |
| | | 83.3 | 10.0 | 6.7 | 100.0 |
| 5 | Muvaris | 31 | 5 | Nil | 36 |
| | | 86.1 | 13.9 | | 100.0 |
| 6 | Vaniyas | 60 | 12 | 3 | 75 |
| | | 80.0 | 16.0 | 4.0 | 100.0 |
| 7 | Tiyyas | 171 | 45 | 19 | 235 |
| | | 72.8 | 19.1 | 8.1 | 100.0 |
| 8 | Artisan and Service Castes | 54 | 7 | 5 | 66 |
| | | 81.8 | 10.6 | 7.6 | 100.0 |
| 9 | Harijans | 43 | 3 | 1 | 47 |
| | | 91.5 | 6.4 | 2.1 | 100.0 |
| 10 | Mappilas | 23 | 2 | 1 | 26 |
| | | 88.5 | 7.7 | 3.8 | 100.0 |
| | Total | 605 | 129 | 66 | 800 |
| | | 75.6 | 16.1 | 8.3 | 100.0 |

**Source:** Tabulated from the field-data.

that the distribution of annual income among the Nambutiri households is almost the same as among the Ambalavasis and the Nayars, and that though these three groups have an edge over the remaining three groups, the majority of the households of all the six groups belong to the same income categories. The magnitude of this change will be more clear from a look at the change in the importance of wet land as the main resource to the Nambutiris versus the other five groups.

It is only too obvious from the steep fall in the total wet land owning households among the Nambutiris and the total wet land owned by them that what the Nambutiris now have as a resource from wet land represents only the crumbs of their opulent past. That this is not sufficient to meet even their basic needs of life is evident enough. As against only 15 per cent of their households not owning any wet land earlier, as many as 51 per cent do not own any wet land now. This should mean that all these households are now forced to buy foodgrains from the market for domestic consumption while in the past these were delivered at their door. Even of the present wet land owning households, only about 19 per cent have land holdings large enough to meet all their annual foodgrain requirements. The rest are dependent on markets for varying periods, with nearly two-thirds for more than nine months (Table 7.2).

**Table 7.2** ***Wet Land Owning Households Depending on Foodgrain Markets in the Post-Reforms Period (Households in Number and Percentage)***

| *Sl. No.* | *Caste* | *Not Dependant* | *Up to 4 months* | *5 to 8 months* | *9 to 12 months* | *Total* |
|---|---|---|---|---|---|---|
| 1 | Nambutiris | 7 | 6 | 5 | 19 | 37 |
| | | 18.9 | 16.2 | 13.5 | 51.4 | 100.0 |
| 2 | Ambalavasis | 3 | 3 | 7 | 6 | 19 |
| | | 15.8 | 15.8 | 36.8 | 31.6 | 100.0 |
| 3 | Nayars | 6 | 8 | 8 | 11 | 33 |
| | | 18.2 | 24.2 | 24.2 | 33.3 | 100.0 |
| 4 | Maniyanis | 10 | 30 | 16 | 26 | 82 |
| | | 12.2 | 36.6 | 19.5 | 31.7 | 100.0 |
| 5 | Vaniyas | 4 | 4 | 7 | 28 | 43 |
| | | 9.3 | 9.3 | 16.3 | 65.1 | 100.0 |
| 6 | Tiyyas | 22 | 13 | 26 | 74 | 135 |
| | | 16.3 | 9.6 | 19.3 | 54.8 | 100.0 |
| | Total | 52 | 64 | 69 | 164 | 349 |
| | | 14.9 | 18.3 | 19.8 | 47.0 | 100.0 |

**Source:** Tabulated from the field-data.

No doubt, despite the significant rise in the total wet land owning households among the remaining five groups and the total wet land owned by them, the extent of market dependence by their wet land owning households is also almost the same as of the wet land owning households among the Nambutiris (Table 7.2). This only points to the fact that the implementation of land reforms has not ushered in an era of plenty or self-sufficiency even among the major beneficiary groups. What is however, more relevant to note here is that their present economic condition is expected to be better than what it might have been earlier. For, earlier, they were forced to surrender a major part of the produce of their leased-in holdings to their *janmis* and this might have trussed them up into an economic straitjacket, whereas now they have the full ownership of these holdings and they do not have to surrender any portion of the produce to anyone.

Despite the fact that the Nambutiris have lost most of their lands, as already seen, the mean size of their holdings is still larger than that of most other groups. Considering that the village economy is still predominantly agrarian, this ought to have given them a dominant status in the post-reforms village economy as well. However, this has not happened. The main reason for this seems to be that the income which the Nambutiris now derive from their lands is not at all commensurate with the size of holdings. Among the Nambutiris about 31 per cent of the households do not get any income from land 61 per cent get only less than Rs. 1,000, and only 8 per cen get Rs. 1,000 or more as annual income from wet and garder lands. Among the remaining five groups, the corresponding percentages are between 14 and 23; 34 and 61; and, 21 and 44 respectively (Table 7.3).

In this connection it is necessary to note that the post-reforms agrarian scene in the village is marked by the absence of share-cropping and tenancy—the twin evils of the land system in Kerala in general and Malabar in particular right up to the beginning of the implementation of the 1969 Act.[4] In such a situation, land has to be cultivated either with family labour or hired labour or both. But, unlike the other five groups, especially the Maniyanis, Vaniyas, and the Tiyyas, the Nambutiris do not have a tradition of cultivation. This has forced them to depend on hired labour for most of their agricultural operations, and placed them at a disadvantage over the other five groups. This is evident from the nature of labour used for wet land cultivation—about 95 per cent of the wet land owning households among the Nambutiris depend on hired labour exclusively whereas only 42 per cent of the households among the Ambalavasis, 52 per cent among

[4] The absence of share-cropping and tenancy was as observed in early 1979. The rise of share-cropping and tenancy in the subsequent years as reported for the larger society might have happened in this village as well. But no attempt was made to collect data on this.

**Table 7.3** ***Households by Annual Income from Land in the Post-Reforms Period (Households in Number and Percentage)***

| *Sl. No.* | *Caste/ Community* | *Nil* | *Up to Rs. 1,000* | *Rs. 1,000 to Rs. 3,000* | *Rs. 3,000 and above* | *Total* |
|---|---|---|---|---|---|---|
| 1 | Nambutiris | 23 | 46 | 6 | — | 75 |
| | | 30.7 | 61.3 | 8.0 | | 100.0 |
| 2 | Ambalavasis | 7 | 11 | 13 | 1 | 32 |
| | | 21.9 | 34.4 | 40.6 | 3.1 | 100.0 |
| 3 | Nayars | 8 | 31 | 17 | 2 | 58 |
| | | 13.8 | 53.4 | 29.3 | 3.4 | 100.0 |
| 4 | Maniyanis | 26 | 82 | 37 | 5 | 150 |
| | | 17.3 | 54.7 | 24.7 | 3.3 | 100.0 |
| 5 | Muvaris | 19 | 17 | — | — | 36 |
| | | 52.8 | 47.2 | | | 100.0 |
| 6 | Vaniyas | 12 | 46 | 15 | 2 | 75 |
| | | 16.0 | 61.3 | 20.0 | 2.7 | 100.0 |
| 7 | Tiyyas | 55 | 132 | 37 | 11 | 235 |
| | | 23.4 | 56.2 | 15.7 | 4.7 | 100.0 |
| 8 | Artisan and Service Castes | 49 | 16 | 1 | — | 66 |
| | | 74.2 | 24.2 | 1.5 | | 100.0 |
| 9 | Harijans | 40 | 2 | 5 | — | 47 |
| | | 85.1 | 4.3 | 10.6 | | 100.0 |
| 10 | Mappilas | 18 | 5 | 3 | — | 26 |
| | | 69.2 | 19.2 | 11.5 | | 100.0 |
| | Total | 257 | 388 | 134 | 21 | 800 |
| | | 32.1 | 48.5 | 16.8 | 2.6 | 100.0 |

**Note:** Dash indicates nil.
**Source:** Tabulated from the field-data.

the Nayars, and between 16 and 23 per cent among the Maniyanis, Vaniyas, and the Tiyyas have such dependence (Table 7.4). The relatively greater dependence of the Ambalavasis and the Nayars presumably reflects their intermediate status in the village society as the high castes ranking immediately below the Nambutiris.

Further, agricultural labourers in this village, forming part of the other major land owning castes of the post-reforms period, were also participants in the numerous peasant struggles. As such they are now reluctant to work for the former landlord class. As a result, they attend to the agricultural operations of the Nambutiris only after exhausting all

**Table 7.4** ***Households by Nature of Labour Used for Wet Land Cultivation in the Post-Reforms Period (Households in Number and Percentage)***

| *Sl. No.* | *Caste* | *FL Exclusively* | *HL Exclusively* | *Mainly FL* | *Total* |
|---|---|---|---|---|---|
| 1 | Nambutiris | — | 35 | 2 | 37 |
| | | | 94.6 | 5.4 | 100.0 |
| 2 | Ambalavasis | — | 8 | 11 | 19 |
| | | | 42.1 | 57.9 | 100.0 |
| 3 | Nayars | — | 17 | 16 | 33 |
| | | | 51.5 | 48.5 | 100.0 |
| 4 | Maniyanis | 9 | 19 | 54 | 82 |
| | | 11.0 | 23.2 | 65.8 | 100.0 |
| 5 | Vaniyas | 1 | 7 | 35 | 43 |
| | | 2.3 | 16.3 | 81.4 | 100.0 |
| 6 | Tiyyas | 17 | 21 | 97 | 135 |
| | | 12.6 | 15.6 | 71.8 | 100.0 |
| | Total | 27 | 107 | 215 | 349 |
| | | 7.7 | 30.7 | 61.6 | 100.0 |

**Note:** Dash indicates nil. FL: family labour; HL: hired labour.
**Source:** Tabulated from the field-data.

other avenues of work, especially after attending to the agricultural operations of the other castes, and that too perfunctorily and at a wage demanded by them. Not infrequently, some of the Nambutiris had left their fields fallow.

If the Nambutiris do not work their wet land with family labour, there is no reason to believe that they work their garden land either, which is larger in size. Among the other five groups garden land is now used for the cultivation of cash crops like pepper, arecanut, and coconut, but among the Nambutiris it is still left mostly uncultivated. As a result, their garden land is also not a major source of income.

The change in the economic status of the Nambutiris was well illustrated by a respondent. When he met one of his former *janmis* standing at a fair-price shop, he asked him sarcastically: 'Once when I brought your *varam*[5] promptly, you returned it asking me to dry it again. Now what will you tell this shop-keeper when he supplies you rotten and stinking rice?'

[5] A *janmi*'s share of the crop due from the tenant.

## Occupational Structure

The occupational structure of the village indicates both continuity and change, while the occupational status of the different groups largely reflects their traditional socio-economic status. Among the Nambutiris, only 29 per cent of the adult population (male and female) are engaged in any occupation, whereas the corresponding percentages for the Ambalavasis, Nayars, Maniyanis, Vaniyas, and the Tiyyas are 37, 39, 52, 48, and 47 respectively. The higher proportion of the unemployed among the Nambutiris seems to indicate their unwillingness to come out of their earlier leisure class status and also their inability to get suitable employment even when they are willing to work.

As may be seen from Table 7.5, of the employed adults among the Nambutiris nearly half (47 per cent) are in relatively high status occupations (sl. nos. 1 & 3); about one-fourth (27 per cent) are in the traditional occupations of priesthood and cooking (sl. nos. 2 & 4); while a few have strayed into cultivation (9 per cent), and a few others (13 per cent) into petty business, white-washing, masonry, and so on. While about 31 per cent of the Ambalavasis and 35 per cent of the Nayars are in relatively high status occupations (sl. nos. 1 and 3), only 11 per cent of the Maniyanis, 16 per cent of the Vaniyas, and 6 per cent of the Tiyyas are in such occupations. In all these five groups however, between 46 per cent and 66 per cent of the employed are in agriculture. That the bulk of those engaged in agriculture among these groups are as agricultural labourers indicates two important aspects of the post-reforms agrarian situation. First, the small and uneconomic holdings, which necessitate not only self-cultivation but also a search for supplemental incomes through hiring out to others as wage labourers. Second, the complicated nature of wet land cultivation, which, despite small holdings and self-cultivation, necessitates dependence on wage labour for at least some of the crucial operations.

Table 7.5 *Households by Occupational Categories (Households in Number and Percentage)*

| *Sl. No.* | *Occupational Categories* | *Nambu-tiris* | *Ambala-vasis* | *Nayars* | *Mani-yanis* | *Muva-ris* | *Vani-yas* | *Tiyyas* | *Artisan/ Service Castes* | *Hari-jans* | *Mappi-las* | *Total* |
|---|---|---|---|---|---|---|---|---|---|---|---|---|
| 1 | Panchayat officer/food inspector/engineer/overseer/bank manager/allopathic doctor/ayurvedic doctor, etc. | 8 | — | 5 | — | — | — | 1 | — | — | — | 14 |
| | | 8.1 | | 4.8 | | | | 0.2 | | | | 0.8 |
| 2 | Priesthood | 25 | — | — | — | — | — | — | — | — | — | 25 |
| | | 25.3 | | | | | | | | | | 1.5 |
| 3 | Teaching/clerical | 39 | 16 | 32 | 33 | — | 25 | 30 | 27 | 5 | 5 | 212 |
| | | 39.4 | 31.4 | 30.5 | 10.8 | | 15.6 | 5.7 | 15.9 | 4.0 | 14.3 | 12.7 |
| 4 | Cooking | 2 | — | — | — | — | — | — | — | — | — | 2 |
| | | 2.0 | | | | | | | | | | 0.1 |
| 5 | Cultivation | 9 | 12 | 15 | 59 | 1 | 22 | 67 | — | 5 | 1 | 191 |
| | | 9.1 | 23.5 | 14.3 | 19.3 | 1.1 | 13.8 | 12.7 | | 4.0 | 2.9 | 11.4 |
| 6 | Agricultural labour | — | 17 | 32 | 143 | 33 | 71 | 258 | 46 | 35 | 1 | 636 |
| | | | 33.3 | 30.5 | 46.9 | 35.5 | 44.4 | 48.8 | 27.1 | 27.8 | 2.9 | 38.0 |
| 7 | Beedi rolling | — | — | 2 | 19 | 9 | 15 | 54 | 30 | 3 | 3 | 135 |
| | | | | 1.9 | 6.2 | 9.7 | 9.4 | 10.2 | 17.6 | 2.4 | 8.6 | 8.1 |
| 8 | Stone-cutting | — | — | 4 | 20 | 35 | 15 | 61 | 4 | 11 | 7 | 157 |
| | | | | 3.8 | 6.6 | 37.6 | 9.4 | 11.5 | 2.4 | 8.7 | 20.0 | 9.4 |

Table 7.5 *Contd.*

| *Sl. No.* | *Occupational Categories* | *Nambu-tiris* | *Ambala-vasis* | *Nayars* | *Mani-yanis* | *Muva-ris* | *Vani-yas* | *Tiyyas* | *Artisan/ Service Castes* | *Hari-jans* | *Mappi-las* | *Total* |
|---|---|---|---|---|---|---|---|---|---|---|---|---|
| 9 | Petty business | 4 | 1 | — | 7 | — | 5 | 9 | 8 | 1 | 11 | 46 |
| | | 4.0 | 2.0 | | 2.3 | | 3.1 | 1.7 | 4.7 | 0.8 | 31.4 | 2.7 |
| 10 | White-washing, masonry, | 9 | 3 | 12 | 17 | 15 | 7 | 43 | 51 | 66 | 3 | 226 |
| | weaving, etc. | 9.1 | 5.9 | 11.4 | 5.6 | 16.1 | 4.4 | 8.1 | 30.0 | 52.4 | 8.6 | 13.5 |
| 11 | Unspecified | 3 | 2 | 3 | 7 | — | — | 6 | 4 | — | 4 | 29 |
| | | 3.0 | 3.9 | 2.9 | 2.3 | | | 1.1 | 2.4 | | 11.4 | 1.7 |
| | Total | 99 | 51 | 105 | 305 | 93 | 160 | 529 | 170 | 126 | 35 | 1,673 |
| | | 100.0 | 100.0 | 100.0 | 100.0 | 100.0 | 100.0 | 100.0 | 100.0 | 100.0 | 100.0 | 100.0 |

**Note:** Dash indicates nil.
**Source:** Tabulated from the field-data.

## Household Composition

Considering that the superior land rights of both the Nambutiris and the Nayars had protected and perpetuated their traditional family organisation, abolition of these rights would have entailed drastic changes in the family organisation. The present household composition among these groups would also lend credence to such a view. Among both the groups, more than two-thirds of the households are of the nuclear type, and the few households which are of the joint type (Table 7.6) are only truncated forms of the traditional family organisation. But, in point of fact, these changes cannot be attributed to the recent land reforms to any significant extent. On the contrary, these were part of the general changes taking place in Kerala society since the early part of this century.

Despite the earlier changes, however, it needs to be emphasised that the introduction of land reforms since 1957 has certainly accelerated the process of change. In the course of field investigation it was found that of the four *illams* which were partitioned since 1970, two had no reason for partition other than fear of losing the remaining land. Of these one was the wealthiest in the village in the pre-reforms period, but after the introduction of the 1969 Act the land was partitioned among the three brothers. In the other two cases partition was primarily due to poverty and family feuds among women, and the realisation that a joint household is not a viable family unit in the present socio-economic circumstances. Similarly, two large *taravads* of the Nayars which split up into nuclear households along with partition of property among the members had no reason for partition other than the fear of ceiling. Even in the case of families which did not own large holdings, partition had taken place due to the fear of losing whatever little land they had owned.

Since ownership of large land holdings by joint families was a characteristic of the traditional family organisation, it may be thought that the existing joint families have been able to sustain themselves because of large land holdings. This is, however, not borne out by facts, as may be seen from Table 7.7. If the surviving joint families are not supported by large land resources, which was a precondition for the existence of

**Table 7.6** *Household Composition in the Post-Reforms Period (Households in Number and Percentage)*

| *Sl. No.* | *Caste/ Community* | *Nuclear* | *Patrilineal Joint* | *Matrilineal Joint* | *Patrilineal Collateral* | *Matrilineal Collateral* | *Combination of 6 & 7* | *Transi-tional* | *Matri-patrilineal* | *Total* |
|---|---|---|---|---|---|---|---|---|---|---|
| 1 | 2 | 3 | 4 | 5 | 6 | 7 | 8 | 9 | 10 | 11 |
| 1 | Nambutiris | 51 | 16 | — | 6 | — | — | 1 | 1 | 75 |
| | | 68.0 | 21.3 | | 8.0 | | | 1.3 | 1.3 | 100.0 |
| 2 | Ambalavasis | 22 | 1 | 1 | 1 | — | — | 5 | 2 | 32 |
| | | 68.8 | 3.1 | 3.1 | 3.1 | | | 15.6 | 6.3 | 100.0 |
| 3 | Nayars | 39 | 2 | 1 | — | — | 3 | 4 | 9 | 58 |
| | | 67.2 | 3.4 | 1.7 | | | 5.2 | 6.9 | 15.5 | 100.0 |
| 4 | Maniyanis | 103 | 21 | 2 | — | 2 | 1 | 14 | 7 | 150 |
| | | 68.7 | 14.0 | 1.3 | | 1.3 | 0.7 | 9.3 | 4.7 | 100.0 |
| 5 | Muvaris | 19 | 9 | — | — | 1 | — | 4 | 3 | 36 |
| | | 52.8 | 25.0 | | | 2.8 | | 11.1 | 8.3 | 100.0 |
| 6 | Vaniyas | 47 | 10 | 1 | — | 1 | 1 | 4 | 11 | 75 |
| | | 62.7 | 13.3 | 1.3 | | 1.3 | 1.3 | 5.3 | 14.7 | 100.0 |
| 7 | Tiyyas | 162 | 22 | 7 | — | 3 | 9 | 13 | 19 | 235 |
| | | 68.9 | 9.4 | 3.0 | | 1.3 | 3.8 | 5.5 | 8.1 | 100.0 |
| 8 | Artisan and Service Castes | 41 | 8 | 1 | — | 3 | — | 8 | 5 | 66 |
| | | 62.1 | 12.1 | 1.5 | | 4.6 | | 12.1 | 7.6 | 100.0 |

Table 7.6 *Contd.*

| *Sl. No.* | *Caste/ Community* | *Nuclear* | *Patrilineal Joint* | *Matrilineal Joint* | *Patrilineal Collateral* | *Matrilineal Collateral* | *Combination of 6 & 7* | *Transi-tional* | *Matri-patrilineal* | *Total* |
|---|---|---|---|---|---|---|---|---|---|---|
| 1 | 2 | 3 | 4 | 5 | 6 | 7 | 8 | 9 | 10 | 11 |
| 9 | Harijans | 24 | 9 | 3 | — | — | 3 | 5 | 3 | 47 |
| | | 51.1 | 19.1 | 6.4 | | | 6.4 | 10.6 | 6.4 | 100.0 |
| 10 | Mappilas | 9 | 1 | 2 | — | — | 2 | 5 | 7 | 26 |
| | | 34.6 | 3.8 | 7.7 | | | 7.7 | 19.2 | 26.9 | 100.0 |
| | Total | 517 | 99 | 18 | 7 | 10 | 19 | 63 | 67 | 800 |
| | | 64.6 | 12.4 | 2.2 | 0.9 | 1.2 | 2.4 | 7.9 | 8.4 | 100.0 |

**Note:** Dash indicates nil.
**Source:** Tabulated from the field-data.

Table 7.7 *Household Types by Land Ownership Categories Among Nambutiris in the Post-Reforms Period*

| *Land Ownership Categories* | *Nuclear* | *Patrilineal Joint* | *Patrilineal Collateral* | *Transitional* | *Matri-patrilineal* | *Total* |
|---|---|---|---|---|---|---|
| Up to 2.00 | 78.3 | 13.0 | 8.7 | — | — | 100.0 |
| | 18 | 3 | 2 | | | 23 |
| | 35.3 | 18.7 | 33.3 | | | 30.7 |
| 2.01–4.00 | 47.4 | 42.1 | — | 5.3 | 5.3 | 100.0 |
| | 9 | 8 | | 1 | 1 | 19 |
| | 17.6 | 50.0 | | 100.0 | 100.0 | 25.3 |
| 4.01–6.00 | 60.0 | 30.0 | 10.0 | — | — | 100.0 |
| | 6 | 3 | 1 | | | 10 |
| | 11.8 | 18.7 | 16.7 | | | 13.3 |
| 6.01–10.00 | 73.3 | 6.7 | 20.0 | — | — | 100.0 |
| | 11 | 1 | 3 | | | 15 |
| | 21.6 | 6.2 | 50.0 | | | 20.0 |
| 10.01–14.00 | 100.0 | — | — | — | — | 100.0 |
| | 3 | | | | | 3 |
| | 5.9 | | | | | 4.0 |
| 14.00 & above | 80.0 | 20.0 | — | — | — | 100.0 |
| | 4 | 1 | | | | 5 |
| | 7.8 | 6.2 | | | | 6.7 |
| Total | 68.0 | 21.3 | 8.0 | 1.3 | 1.3 | 100.0 |
| | 51 | 16 | 6 | 1 | 1 | 75 |
| | 100.0 | 100.0 | 100.0 | 100.0 | 100.0 | 100.0 |

**Note:** Dash indicates nil; households in number and percentage; land ownership categories in acres.
**Source:** Tabulated from the field-data.

joint families in the past, it would appear only too obvious that joint families are no longer a stable type, are in a process of transition, and, in course of time, are likely to give way to the nuclear family which has already become the dominant type.

Next to the Nambutiris and the Nayars, the only major group which followed the joint family system was the Tiyyas. In fact, the traditional family organisation of the Tiyyas was modelled on that of the Nayars, based on the matrilineal system of inheritance and descent (*marumakkattayam*) and the duolocal system of residence. However, as a majority of the Tiyyas belonged to the labouring class and the group as a whole had a very low socio-economic status, the family organisation among them was not as stable as among the Nayars. Compared to the large number of stable and well-knit Nayar joint families (*taravads*) in the traditional Kerala society, there were only very few Tiyya *taravads*, so much so that among the Tiyyas the matrilineal principles operated not so much in actual social life, nature of inheritance, and membership composition of the households as in the minds of the family members. Accordingly, land reforms which have been instrumental in accelerating the process of family change among the Nambutiris and the Nayars might not have had any major impact on the Tiyya family, except in the case of the few households which might have owned large holdings in the pre-reforms period.

## Religious Practices

Unlike the other aspects of the village society, changes in religious practices are mainly in relation to the Nambutiris. This has to be explained by their traditional social status as a caste of the highest ritual ranking which made religious pursuits far more significant to them than to the other groups, and which coupled with their land right as *janmis* also enabled them to dominate the religious life of the other groups.

In fact, as a caste of the highest ritual ranking, practically every act in life is religious to the traditional Nambutiris. Broadly, their religious practices include sixteen *rites de passage* (*shodasa kriyas* or *shodasa samskaras*), performed for and by the individual family members,[6] the daily worship (*thevaram*), and the seasonal ceremonies performed to propitiate the family deities.[7] There is no *illam* in the village without a *padinhatta*,[8] and four of the *illams* have shrines of Aiyappan, Vishnu, Subramanian and Devi, respectively. Daily worship and seasonal festivals performed in these shrines are of importance to the whole village. In the pre-reforms period, a feast in the *illams* was a daily affair at which apart from feeding the Brahmins and offering gifts to them food was served to the tenants and other dependents. It is generally remarked in Kerala that the Nambutiris did not mind spending all they had for a feast, festival, or a Kathakali performance.[9]

Although the Nambutiris traditionally enjoyed ritual superiority over the other social groups, it is important to note that it was because of their land ownership that they could remain a leisure class devoted to the pursuit of learning and religious practices, supported and sustained by the efforts of the other groups. But, in the post-reforms period, they cannot afford to pursue their religious life with such abandon. Most of the important life-ceremonies are not performed by them with the pomp and festivity of the pre-reforms period. The fact that in this village most of these ceremonies were meticulously and religiously followed in the pre-reforms period, while in some other villages these had been abandoned even earlier apparently owing to the impact of modern education and such other factors, seems to suggest that the post-reforms changes are evidently caused by the impoverishment of the

[6] For a discussion of these rites, see Rao, 1957:112–15.

[7] Namboodiripad's description of his *illam* gives some idea of the importance of family deities among the Nambutiris. He noted that gods and goddesses were inseperable from the daily life in the *illam*, that 'proper arrangements existed in the *illam* to tend them, to render them prompt service and to keep them in good humour' and that 'there were persons specially assigned with the responsibility of looking after them carefully'. See, Namboodiripad, 1976:21–28.

[8] The western chamber of the house, usually meant for worship.

[9] A case in point is Kunjan Pilla's illustration that though the Nambutiris insisted on having five hundred *valsans* (a cake made from jack-tree leaf) for the Ganapati Homam, 90 per cent of these were enjoyed by their tenants and neighbours and not by members of their household. See Pilla, 1966:87. No doubt, often this munificence was at the cost of their tenants.

Nambutiris as a result of the land reforms Act. The Aiyappan temple attached to one of the *illams* was closed for two years without even the daily offerings. To avoid its total ruin, a public committee constituted by the villagers with members from different castes took over its management. The committee spent about Rs. 2,000, collected from the villagers, for renovation and consecration (*kalasattam*) of the neglected shrine. The fate of the other shrines is no different. However, daily offerings continued to be made in them, and since they are not as important as the Aiyappan temple, the villagers have not taken over their management.

In one of the *illams* only an imitation Teyyam devoid of costumes (*vellattam*) is performed in the post-reforms period. The head of the *illam* was hopeful that his family deities would give him the means to perform at least this without fail.[10] In two other *illams* only a nominal ritual is performed on the day of the Teyyams. Only in one *illam* are Teyyams still performed in the traditional way.

Upper castes domination over the religious life of the subordinate groups was not unique to the Nambutiris, but was an attribute of the institution of landlordism as stated by Logan and as noted in Chapter 2. But since the *janmis* in Kodakkad were mostly Nambutiris, it was they who dominated its religious life. Their dominance was manifested in two distinct ways. The first was their insistence that the subordinate groups adhere to the traditional caste practices marked by the latter's subordination to the Nambutiris, and to the high castes in general, an insistence which lingered on until the implementation of the land reforms Act, especially in relation to the lower castes. The second was the organisation of the subordinate groups into different cult-centres controlled by the *janmis*. The custom of seeking the consent of the *janmis* for important religious ceremonies, especially

[10] Teyyams are costly ceremonies. By the present village standards the cost of having a Teyyam performed may be anywhere between Rs. 300 and Rs. 1,500, depending on the nature of the Teyyam. In the case of the Nambutiris, the cost will be much more as most of them worship more than one Teyyam and as in the past at the time of the performance the *illams* are expected to arrange a feast, especially for the dancers and their kith and kin.

the Teyyams, performed in the households of these groups, the power of the *janmis* to make or mar these ceremonies and to punish the dancers, and the obligation of the cult-centres were all explicit. In fact, in the pre-reforms period many of the religious ceremonies among these groups were performed only at the pleasure of the *janmis*, and there were several instances of these ceremonies, especially the Teyyams, being stopped abruptly by the *janmis* for some reason or the other. But the abolition of landlordism and tenancy and conferment of ownership rights on the cultivating tenants of the lands leased in by them have put an end to such religious domination. In the post-reforms period, the Nambutiris do not exert any influence at all over the religious life of the other groups.

In the pre-reforms period the Nambutiris could have also dominated the community temples of the subordinate groups. However, this did not happen because of the fact that the temples of the Nambutiris of this village as also of several other villages in north Malabar are of individual families and not of the community as such. In the absence of a community temple of the Nambutiris, according to tradition, it was the community temple of the Nayars, situated in a nearby village, which dominated the community temples of the other groups. This temple, the most important of the Nayar temples in the whole of the northern part of north Malabar, was traditionally owned and managed by a dominant Nayar *taravad*. But, because of the failure of this *taravad* to manage its affairs properly, in the 1950s it was taken over by the government and its management was entrusted to a public trustee. However, as a vestige of tradition, the head of this *taravad* continues to be the overlord of the temple, and for any important ceremony or festival in it his permission is still considered necessary by the Nayars of the area. It is to this temple that the Nayars of the village belong for religious purposes. The head of the Nayar *taravad* who has overlordship in this temple traditionally enjoys such a right in the Maniyani temple situated within the village, in the Vaniya temple situated in a neighbouring village to which the Vaniyas of this village as also of a number of other villages belong, and in the two Tiyya temples, one situated within the village and the other situated in another village.[11] For the grand Teyyam festival (*perum-kaliyattam*) conducted in the Vaniya temple for about a week in every twelve years, not only is the sanction of the head of the Nayar *taravad* necessary, but water, fire, rice, and such other items are also required to be brought from the Nayar temple.

The religious dependence of the Maniyanis, Vaniyas, and the Tiyyas on the Nayar temple is reminiscent of the traditional village social hierarchy in which they were inferior to the Nayars. It also reflects the hierarchy of the deities in the region, for the deities of the Nayar temple are believed to be superior to those of these three groups. In the pre-reforms period, this religious dependence also reflected the economic dependence of these groups on the Nambutiris and the Nayar *janmis*, when the dependent groups took great care to adhere to the traditional religious norms, with a view to avoiding the displeasure of their *janmis*. Although the religious dependence still continues, in the absence of the pre-reforms economic dependence it is again merely a vestige of tradition.

While on changes in religious practices, it needs to be mentioned that in the religious life of both the non-polluting and the polluting castes of north Malabar the Tiyyas also have certain explicit functions, though quite unlike the Nambutiris these functions do not in any way indicate their religious importance. On the contrary, they are manifestations of the customary patron-client relationship and the extra-economic coercion of the landed castes on their dependents. The Tiyyas are closely associated with the Teyyam cult. Some of the important Teyyams such as *kathivanur-veeran*, *kunhikoran* (part of the *vayanattukulavan* Teyyam), and *palanthayi-kannan* (the hero of *vishnumurti* Teyyam) are woven around Tiyya heroes; in some of the important Teyyams such as *kathivanur-veeran*, a Tiyya is obliged to act as the oracle (*komaram*); in some others such as *muthappan*, a Tiyya on vow has to be the mediator (*nottirikkunna-aal*)

[11] The Ambalavasis do not have a community temple of their own in or around the village. This has to be explained by their traditional occupation as temple functionaries of the high castes because of which there might not have been any need for them to have their own temples. Moreover, they are very few in number.

between the dancer and the deity; in yet a few others which require toddy, a Tiyya has to carry it. The Tiyyas also have a hereditary obligation in limited areas to supply palm-leaf torches for the Teyyam dances. Similarly, if a death occurs in a high caste house, a Tiyya has to cut down the tree to be used for cremation. All such services are customarily rewarded.

Apart from the above religious functions, the social life of the Tiyya community itself is centred round religion. The Tiyyas of a given area belong to a temple organisation (*kazhakam*), and male members of the community belong to it in various capacities as bachelors (*ara-valiyakkar*), married persons (*valiyakkar*), temple functionaries, office-bearers, and so on. Traditionally the members have to conform to the religious rules of the temples by helping in carrying out the ceremonies, festivals, conforming to the caste-rules, and so on.

The biggest temple in Kodakkad, both in size and importance, is the *mappidacheri* temple of the Tiyyas. According to some of the Tiyyas of the village, this temple originally belonged to the Nambutiris. It was gifted to a Nambi, who, unable to manage it, gifted it to the Tiyyas. It is now managed by seven right-holders (*avakasikal*). The first and most important right is of the Nambi himself. He has the supervisory authority (*koyma*) and it is obligatory to obtain his permission for performing any important functions in the temple. The other right-holders in their order of precedence are the *anthithiriyan*, *embron*, *kalasakkaran*, *kodakkaran*, *achan*, and *valiyakkar*.[12] It is held that one of the Nayar *taravads* had the second important right, called *akambadi* but, for purposes of management of the temple, this was entrusted to the *anthithiriyan* along with the gift of some land to the temple. Once in five years, this temple conducts a *kaliyattam* lasting for five

[12] *Anthithiriyan* is the person who lights the lamps of the temple. *Embron* is the priest. The term is a corruption of *Embrantiri*, the priest of the Brahmin temple. *Kalasakkaran* is the person who places toddy at the place earmarked for it in the temple (*kalasathara*), and also carries the weapons of the temple. *Kodakkaran* is the umbrella-holder. *Achan* (literaly elder) also takes the weapons of the temple. *Valiyakkar* are the married men of the Tiyya caste who appoint the other functionaries (*pravruthikkar*) of the temple.

days, when Teyyams of *paradevata*, *bhagavati*, *rakteswari*, and *pataveeran* are performed on all the five days. Expenses for the first day are met by the Nambi, and for another day by the Nayar *taravad* which gave its right and land to the temple.

While the subservience of the temple to the Nambi still continues, it is on the belief that the goddess of the temple expects it and will harm the villagers if his importance in the temple is overlooked.

While changes have taken place in the religious life of other groups, such as the Nambutiris and the Nayars (especially the former), and in the religious practices of the Tiyyas in relation to the upper castes, their temple-centred religious life still continues as in the past, and land reforms have not affected it significantly. If anything, land reforms have only given them more freedom in their religious pursuits, as they now do not have to depend on their *janmis* who traditionally had the supreme authority (*melkoyama* and the supervisory authority of the Tiyya temples. Further, it has been found that religion among the lower castes has of late gained a new lease of life, and as a result. Teyyams have also gained more importance, especially among the Tiyyas.

## Power Structure

With their economic dominance and ritual superiority, in the pre-reforms period the Nambutiris wielded enormous power and authority over the entire village, and dominated the village power structure. Until as recently as a decade before the abolition of landlordism, the biggest *janmi* of the village acted as a sort of uncrowned king. His *illam* served as a court to decide village disputes[13] of all kinds and his word had the force of law. Important government officials who visited the village usually camped at his *illam*. He was invariably consulted by all public officials before they interfered with the

[13] In settling village disputes, the punishments were generally *vazhipadu* (that is, offering to the *dayaramangalam* temple of the Nayars in the neighbouring village for the non-polluting castes, and to the *mappidacheri* temple of the Tiyyas in the case of the polluting castes).

village affairs.[14] The other *janmis* also had their importance in the village affairs of their respective areas, especially in relation to the tenant households. This situation has completely changed in the post-reforms period. With the introduction of panchayati raj and adult franchise, votes began to yield more power than birth and other ascriptive criteria, and with this the Nambutiris as a numerical minority began to lose out most of their importance in village affairs to the castes which could marshal large numbers. However, so long as they remained dominant in the land system, they could retain some of their importance, if not in the context of the whole village at least in relation to their tenants and other dependents. But with the implementation of the land reforms Act they have lost even this importance. Of the remaining groups only the Nayars and the Maniyanis had some importance in village affairs, and were not far removed from the locus of power. However, compared to the Nambutiris these two groups were far less important in the power structure.[15]

In the post-reforms period, the *Karshaka Sangham* (KS), the village panchayat, and the CPI(M) are the important organs through which the power structure of the village is manifested. Of these, the KS is an organ of the CPI(M), and the only peasant organisation in the village. It stands to protect the interests of the cultivators and its membership is mainly drawn from the wet land owning households. To the peasants of the village it is of immense importance in various matters relating to their agricultural operations (such as settling disputes, and getting loans and pumpsets). If anyone proves recalcitrant to it that would mean inviting trouble in carrying out the agricultural operations smoothly. To avoid

[14] Probably his importance in village life would have continued right up to the introduction of the land reforms, but his death in a road accident put an abrupt end to it.

[15] The head of one of the Nayar *taravads* and of one of the Maniyani *taravads* situated in two different parts of the village exercised some influence in village affairs in the areas around their *taravads*. The Nayar *taravad* lost its influence in the village not so much as a result of loss of land, thanks to the rearrangement of holdings by its head, but as a result of the introduction of panchayati raj and adult franchise, and the spread of the communist movement. In the case of the Maniyani *taravad*, it was the abolition of the hereditary headmanship in 1961 which resulted in its loss of influence.

this, most of the households pay a membership fee, which is of course very nominal. Among the wet land owning households of the Nambutiris none is a member of the KS whereas between 15 per cent and 49 per cent of the other five groups (26 per cent, 15 per cent, 39 per cent, 44 per cent and 48 per cent respectively among the Ambalavasis, Nayars, Maniyanis, Vaniyas, and the Tiyyas) are its members.

The village panchayat comprises this village and two other neighbouring villages of Pilicode and Maniyat, and is situated in Pilicode. It consists of seven members (six elected and one nominated) including one president and one vice-president. At the political level all these members belong to the CPI(M), while at the caste level two (including the president) are Nayars, one is a Maniyani, and four (including the vice-president) are Tiyyas.

The CPI(M) is the dominant political party of the village as it is evident from the political affiliations of the villagers, and the membership composition of the village panchayat, and the KS. But, only about 15 per cent of the Nambutiri households support this party as against between 29 per cent and 75 per cent of the households of the other five groups (Table 7.8).

The membership composition of the KS, village panchayat and the CPI(M) clearly indicate that the Nambutiris not only do not dominate the post-reforms power structure but, more importantly, are also far removed from its locus. As against this, the Tiyyas (who, as a polluting caste and a landless class did not find a place in the power structure in the pre-reforms period) dominate it in the post-reforms period. Thus, there has been a complete reversal in the positions of these two groups.

The dominance of the Tiyyas in the post-reforms power structure is the result of the interplay of some important social forces. As a caste the Tiyyas have numerical preponderance over all other social groups in the village. The abolition of landlordism and the conferment of ownership rights on the cultivating tenants of the lands leased in by them, preceded by the introduction of panchayati raj and adult franchise, have helped them relegate to the background the traditional principles of power and authority relations, and

**Table 7.8** ***Households by Political Affiliation (Households in Number and Percentage)***

| *Sl. No.* | *Caste/Community* | *Political Parties* CPI(M) | *Rightist Parties* | *Total* |
|---|---|---|---|---|
| 1 | Nambutiris | 11 | 64 | 75 |
| | | 14.7 | 85.3 | 100.0 |
| 2 | Ambalavasis | 14 | 18 | 32 |
| | | 43.8 | 56.3 | 100.0 |
| 3 | Nayars | 29 | 29 | 58 |
| | | 50.0 | 50.0 | 100.0 |
| 4 | Maniyanis | 108 | 42 | 150 |
| | | 72.0 | 28.0 | 100.0 |
| 5 | Muvaris | 32 | 4 | 36 |
| | | 88.9 | 11.1 | 100.0 |
| 6 | Vaniyas | 50 | 25 | 75 |
| | | 66.7 | 33.3 | 100.0 |
| 7 | Tiyyas | 176 | 59 | 235 |
| | | 74.9 | 25.1 | 100.0 |
| 8 | Artisan and | 39 | 27 | 66 |
| | Service Castes | 59.1 | 40.9 | 100.0 |
| 9 | Harijans | 9 | 38 | 47 |
| | | 19.1 | 80.9 | 100.0 |
| 10 | Mappilas | 9 | 17 | 26 |
| | | 34.6 | 65.4 | 100.0 |
| | Total | 477 | 323 | 800 |
| | | 59.6 | 40.4 | 100.0 |

**Source:** Tabulated from the field-data.

bring to the fore numerical strength as an important factor in wielding power and influence.

Further, as mentioned earlier, for a long time this village had been in the thick of peasant and communist activism and the Tiyyas of the village had been the most important of the activists. In the 1940s, a few of the relatively better off Tiyyas—also the peasant and communist leaders of the village—founded a cooperative society in the village. As this society extended financial help to the needy villagers its founders could use it as a centre for enrolling members for the communist party, and also as a source from which the Tiyyas as a group could gradually wield power and influence.

Over the years the society has grown in size and stature, and the Tiyyas have throughout dominated its important positions and activities.

Of the remaining groups, the Nayars and the Maniyanis are still not far removed from the locus of power. Some of the Nayars were proud to reveal that as members of the Congress Party they had actively participated in the liberation struggle (*vomochana samaram*) which brought down the first communist ministry, and courted arrest for picketing one of the government offices in a nearby town. In the 1960s, they started a unit of the Nayar Service Society (NSS) in the village, but it soon ceased to exist because of the squabbles among its founders. Activities of this kind have continued into the post-reforms period, and a majority of the Nayars continue to be associated with various rightist parties, especially the Congress (I). However, the existence of the CPI(M) as a power has considerably eclipsed their importance and weakened their activities.

Taking the cue from the NSS and the SNDP Yogam, in the 1970s the Maniyanis also founded a communal organisation, the *Kerala Yadava Sabha*, claiming for themselves the status of Yadavas. This organisation is affiliated to the All-India Yadava Mahasabha in New Delhi, and has been campaigning to get representation to members of the Maniyani community in the various organisations of the state government. Through active lobbying in the state capital by some of the Maniyani leaders of the village belonging to this organisation as well as to the Congress (I), in 1976 a high school was established in the village under the management of a Nayar, also belonging to the Congress (I). Prior to this, in 1972, a government rural dispensary was started in the village mainly through the lobbying of the same leaders.

## Nambutiris and Social Relations

In contrast to their earlier role as the main agents of social control, restricting and regulating the village social relations in such a way as to ensure their socio-economic and religious superiority over the other groups, the Nambutiris now neither act as agents of social control nor exert any similar influence on the other groups. Instead, they are socially and economically dependent on these groups, and this has forced them to adjust and adapt to the changing social scene. One indication of this is the employment in manual occupations by some members of this pre-reforms landlord class which despite changes elsewhere clung to their aristocratic status, spurned labour and led a leisure-class life. As it is evident from the occupational data already discussed, of the employed adults among the Nambutiris about 9 per cent are engaged in cultivation, another 9 per cent in white-washing, masonry, and weaving, and about 4 per cent in petty business. Those employed as weavers were found to be three girls from a poor *illam*, who commuted daily to the work-place in a nearby town.[16] It was also found that two of the secondary school educated boys from another *illam* (which was the second wealthiest in the pre-reforms period), tried their hand at stone-cutting for some time. Similarly, a middle-aged Nambutiri from a third poor *illam* was found serving food daily, on a payment basis, to a group of Maniyani and Tiyya teachers, who of course maintained their own plates and tumblers.

In another instance, a Nambutiri youth working as teacher in the government primary school of the village headed by a Harijan (Pulaya) refused to mention the caste of his two sons in the form for admission to the school. When the headmaster asked him why he did not mention their caste, he expressed his desire to bring up his sons in a secular way, and the former was very pleased about the good sense of the Nambutiri. But he felt annoyed when he came to know that the Nambutiri had subsequently performed the *upanayanam* of these boys. The Nambutiri was however frank enough to admit that he could not afford to displease either the headmaster with whom he was working, or his community which was still traditional.

Instances like these, and instances of the dependence of the Nambutiri community on other groups for various

[16] It needs to be mentioned that traditionally the Nambutiris and other high castes had looked down upon weaving as an impure occupation. In north Malabar it continues to be the caste occupation of the Chaliyas.

purposes (such as management of their temples) reflect a new pattern of social relations cutting across traditional caste barriers. They also reflect a reversal of the traditional social roles. Some of the Nambutiris now play the role of dependents, while members of the traditionally lower-ranking social groups for whom they work or on whom they depend play the role of patrons. Such instances are also symptomatic of further changes in the social relations of the Nambutiris vis-a-vis the other social groups.

### *Kerala Nambutiri Samajam*

The Nambutiri community as a whole has also realised the need for change, though somewhat belatedly. This is evident from the working of an association, the *Kerala Nambutiri Samajam*. As its activities symbolise further changes within the community, it will be relevant to note them briefly. The *Samajam* was formed in 1969 as a registered society by a few young and educated Nambutiris, mostly school teachers of Kodakkad and neighbouring villages. It has to be underscored that the initiative for change has come from the young and educated Nambutiris. This was so in the case of the earlier reforms movement also. In 1969 itself the *Samajam* established 22 of its units in Cannanore district, with an office at Payyanur. In Kodakkad it established two such units with a membership of 92 Nambutiri households in one and 67 in the other, representing a number of neighbouring villages as well.

The *Samajam* sought to wake up the Nambutiris to the realities of the present and draw them into the mainstream of social life of the present. One of its important activities was a seven-day *sramdan* (free manual labour campaign), organised in 1970. This campaign, begun with the chanting of a Vedic hymn to mean 'let this effort of ours to weed out false pride and prestige that have cancerously grown into our souls be a success,' was reported to be a touching scene.[17]

[17] This was reported at length on the front page of one of the Malayalam newspapers with a photograph of the Nambutiris in action. See *Mathrubhumi*, 25 October 1970. Welcoming this campaign V.T. Bhattathirippad, one of the leaders of the earlier Nambutiri reforms movement wrote a lengthy letter in the same daily. See *Mathrubhumi*, 4 November 1970. Another daily, wrote a lengthy editorial (see *Kerala Kaumudi* 4 November 1970).

To start with, about 40 Nambutiri men and women toiled with sickles and shovels to clear the shrubs and weeds from a small rubber estate of a Nambutiri in a village close to Kodakkad. This was followed by the harvesting of paddy in about two acres of land owned by another Nambutiri, and the cleaning and repairing of a pond owned by a third Nambutiri. In the same year, the *Samajam* sent a delegation to the state capital to submit a memorandum to the government requesting amendment to the land reforms Act enhancing the compensation payable to the dispossessed *janmis*. As part of the campaign an elderly Nambutiri went on *satyagraha* in front of the Secretariat.

With a view to strengthening its economic base and helping the poor Nambutiris, the *Samajam* started a cloth shop at Payyanur in 1971 investing about Rs. 25,000. The shop did good business for about three years, but unexpectedly, one of its partners decamped with money, resulting in its closure. Among its other activities, the *Samajam* acted as a centre for training the Nambutiris. At its instance, one Nambutiri was 'trained' as an agricultural labourer, another as a mason, and a third was helped to build a small tiled house, solely with the free labour of the *Samajam* members. There were also a few instances of Nambutiri boys practising stone-cutting with the help of the members of the Muvari community.

With a view to finding bridegrooms for the numerous Nambutiri spinsters, it started a matrimonial bureau at Payyanur. Initially there was a good inflow of applications, horoscopes and photographs, and the *Samajam* successfully negotiated four alliances. But the bureau could not do much work as the enthusiasm of the applicants gradually diminished.

The *Samajam* has been of the view that in the present social context a joint household or joint ownership of property is disadvantageous. Accordingly, it has been encouraging partition of joint property and setting up of nuclear households. At its instance, in 1974, the properties of three *illams* were partitioned along with splitting up of the households.

In 1975 the *Samajam* organised a two day Nambutiri conference at Mathamangalam, about 8 kms east of Payyanur town. About 800 Nambutiri men and women from different parts of Kerala participated in it. The conference resolved to appeal to the government to abolish caste-based reservations in government service and educational institutions and to adopt income-based reservations; and offer protection to the former small and marginal *janmis*. The conference was attended by some of the activists of the earlier *Nambutiri Yogakshema Mahasabha*. Taking the cue from this conference, the *Sabha* was subsequently revived as it had been inactive for a long time.

In May 1979, the *Samajam* arranged a group-*upanayanam* and *sam-aavartanam* for four Nambutiri boys, with the aim of reducing the expenditure on such religious practices. As a further measure of economy it tried to be content with only one priest and one *homam* for all the four boys; but the priest engaged and the parents of the boys were not ready for such a drastic change in the ceremonies, and insisted on having a separate *homam* and priest for each of the boys.

The *Samajam* also had other activities like providing schoolbooks to the Nambutiri boys, scholarships to selected students, and cloths and umbrellas to the Nambutiri widows. However, its activities have not met with the approval of all the Nambutiris. While the *Samajam* continues to exist it is not as active as it was a few years ago. Not much help or cooperation is forthcoming from the members and the same few Nambutiris who formed it continue the crusade all alone. Since most of them are full-time teachers they do not find much time for social service.

## Beneficiary Groups and Social Relations

Earlier, all the five beneficiary groups were inferior to the Nambutiris, both socially and economically, as castes ranking below them and as cultivating tenants. Therefore, their relation to the Nambutiris was one of socio-economic subservience. Since the caste-based social subservience was reinforced by economic subservience, the former lingered on so long as the latter persisted. However, the gain of land by these groups as a result of the implementation of the land reforms Act has put an end to both. Consequently, now the relationship of these groups with the Nambutiris is one of equality rather than subordination.

The magnitude of this change is however greater in the case of the Tiyyas. For the Tiyyas traditionally belonged to a 'culture of deprivation'. In fact, as a polluting caste, they were socially subservient to all the non-polluting castes. This subservience was expressed through numerous servile social practices based on the caste system and the caste-based institution of landlordism. But consequent on the weakening of the caste system in the whole of Kerala, this subservience was weakened in their relation towards all but their *janmis*, to whom they were forced to remain subservient because of their economic dependence on them. The following observations by Gough will illustrate this point:

> In spite of the various political and social changes, the traditional relationship of servitude persisted in a modified form in rural areas of Kerala . . . . This occurred where a dominant caste retained possession of most of the land, or where land was still used primarily for subsistence cultivation, and subordinate castes can find no superior employment outside their traditional services. Where population pressure and movement had not forced a complete breakdown of hereditary services, and there was no replacement of a labour market, caste still held some power. In these conditions the dominant caste can use its economic power to enforce observance of traditional inter-caste laws. (Gough, 1959:120)

To cite an example, in pre-reforms Kerala society if the *janmi* belonged to a high caste and the tenant to a polluting caste, the latter was expected to deliver to the former his share of crop through persons of a non-polluting caste. The practice lingered on until the land reforms Act came into force. One of the Tiyya respondents from the village informed that just a year before the enforcement of the Act he refused to adhere to this caste practice. When the Nambutiri *janmi* insisted on his getting persons of non-polluting castes to carry the paddy to his granary, he threatened to take it away back home. He confessed that he would not have dared to do this in the pre-reforms period for fear of harassment and eviction, but the Stay of Eviction Proceedings Act of 1967 and the land reforms Act had rendered such fear unnecessary. This only confirms Aiyappan's observation that 'without some degree of economic freedom the underprivileged communities are seldom able to take advantage of the social legislations conferring new privileges on them' (Aiyappan, 1965:10).

While the Tiyyas have developed a new pattern of egalitarian relationship with the castes above them in the hierarchy, the same cannot be said about their relationship with the castes below them. If anything, the improvement in their economic condition has served only to accentuate the social distance between them and the castes lower in the hierarchy. This view was supported by a number of respondents belonging to the Nambutiri, Nayar and the Harijan communities. One of the Harijan respondents, headmaster of the government primary school in the village and a political leader, narrated an interesting incident to support that the Tiyyas are still caste conscious and behave towards the lower polluting castes almost the same way as the Nambutiris had done towards the Tiyyas in the pre-reforms period.

Of the two religious centres to which the Tiyyas of the village belong—viz., nearly half to the *mappidacheri* temple in their own locality in the village and the rest to the *vanillam kazhakam* in an adjoining village—the former has a Teyyam festival (*kaliyattam*) conducted once in three years. As part of this festival, a Teyyam is taken out by the temple authorities in a procession to the households of all except the Harijans of the temple area. In 1973 when this practice was repeated, the Harijan respondent resented it and requested the temple authorities not to single out the Harijans. When his request went unheeded, he took it as an affront and reported the matter to the Removal of Untouchability Squad. The temple functionaries (including the Teyyam dancers) were arrested and produced before the law court. The litigation went on for about two years and at the end the accused were acquitted after they tendered an apology for the offence.

The foregoing analysis of the social consequences of the implementation of the KLRAA of 1969 clearly shows that in Kodakkad the loss of land by the group in which land was traditionally concentrated along with status, power and authority, and the gain of land by the groups which were traditionally dependent on and subservient to it have resulted in the loss of status, power and authority by the former, gain of socio-economic status by the latter, and alteration in the traditional hierarchical socio-economic relations into egalitarian relations.

The implementation of land reforms has not brought about any significant change among the Muvaris, artisan and service castes, Harijans and the Mappilas. This is owing to the fact that these groups have not benefited much by the land reforms Act, especially from the redistribution of wet land. As a result of these, as in the pre-reforms period the socio-economic status of these groups continues to be lower than that of the other six groups. This is evident from the distribution of annual income, and the income derived from whatever land they now own. As only very few households among them own any wet land, even in the post-reforms period virtually all their households are totally dependent on the market for their foodgrain requirements. Occupationally also they continue to have a lower status.

Moreover, the Muvaris, and the artisan and service castes were economically dependent on the Nambutiris—Muvari women as household servants and others as washermen, Teyyam dancers and so on. But the deterioration in the economic condition of the Nambutiris has forced them to dispense with the services of these groups, which has affected their economic condition as well.

The economic impoverishment of these groups is not without consequences for the village society. To cite an instance, one of the Harijan Teyyam dancers who was performing the *kattumurti* Teyyam for the Maniyani temple was living in the temple area (*pati*) on a piece of land allotted by a Maniyani *taravad* which supervised the functioning of this temple. On the eve of the implementation of the land reforms Act, the Maniyani *taravad* claimed ownership of this land and gave its occupant only ten cents as per the provisions of the Act. As a protest against this, the Harijan dancer stopped dancing for the Maniyani temple. Instead, he started working as stone-cutter in another village.

To cite another instance, once a year (in the Malayalam month of *Karkitakam*),[18] the Vannan and Malayan Teyyam

[18] The last month of the Malayalam year corresponding to July-August.

dancers of each area in the village had to perform Teyyams for the village as a whole. They went from house to house, the Vannan as Adi (Karkitakatti or Parvati) and the Malayan as Vedan (Siva), either jointly or individually. The belief is that these Teyyams drove away all the evils and diseases of the village, and ushered in health and prosperity. The offerings made by the villagers were adequate for the survival of the dancers for quite some time. However, as a result of the partition of land holdings and splitting up of households in the village, and the reluctance of the villagers to make offerings any more, the dancers cannot make a living out of their traditional work. Hence some of them have given up this practice, and taken to other occupations such as stone-cutting, and beedi-rolling, which fetch them a higher income.

# 8

# Conclusion

Why does the peasantry remain passive victims of an oppressive and exploitative land system, and by implication, of an oppressive and exploitative social order? When and how do they struggle for their legitimate rights and with what success? When their struggles eventually force the state to recognise their rights as legitimate, why does such recognition remain at the level of mere rhetoric? When some peasant-oriented governments attempt to translate this recognition into reality, under what conditions and with what consequences do they do so? These, in brief, are the issues emerging from this study.

The specific context of the study has been the interplay of the historical processes of peasant struggles, land reforms and social change in Malabar. In trying to understand this interplay, the study has at the outset postulated that far from being liberal gifts from enlightened governments, land reforms have been historical processes necessitated by peasant struggles and that any coherent and comprehensive understanding of social change on the basis of such reforms should necessarily be in the context of the related peasant struggles.

In keeping with these postulations the study has attempted

to understand its subject in two related ways. First by documenting the peasant struggles and land reforms in Malabar from the second half of the nineteenth century right up to 1970, against the backdrop of the land and society in pre-British Malabar. Second, by looking at the ultimate outcome of these struggles, namely, the Kerala Land Reforms (Amendment) Act of 1969. As the conclusions emerging from these endeavours are related to the main issues discussed earlier, they should necessarily be in the context of a recapitulation of these issues.

Though the land system in Malabar was highly oppressive and exploitative even in its pre-British pristine form, it was only long after the advent of British rule that the peasantry began to protest. Their passivity might have been because of the nature of the land system itself. It was caste-based, Brahmin-centred, and upper caste dominated, founded and functioning on an intricate system of social closure which closed the outlets of protest to not only the peasantry but also the various other oppressed and exploited groups.

When the conditions of the peasantry deteriorated further following the Mysorean and British rules, and the peasantry eventually began to protest through their long-drawn out struggles, the course of these struggles and their nature and character were in the context of the prevalent socio-historical and geographic conditions.

Thus, the struggles in the first phase were by the lower Mappila tenantry but not by their Hindu counterparts. The variations in the socio-cultural antecedents of the two principal sections of the oppressed (namely, the Muslims and the Hindus) explain these selective outbursts. The struggles of the Mappilas were the result of a combination of factors—inheritance of a militant tradition, crystallisation of an independent communal identity, and increasing disrespect for traditional Hindu institutions, all precipitated by the Portuguese intrusion; a new sense of euphoria and freedom, new possession of land, and embitterment of relations with the Hindus, all following the Mysorean occupation of Malabar; and finally, decline in socio-economic conditions and further embitterment of relations with the Hindus, consequent on the British land policy. The silence of their Hindu counterparts was because of the tenacity of traditional authority and their continuing submission to it.

The concentration of the struggles in south Malabar was because of the concentration of the Mappila population in that region. The communal idiom of the struggles was contingent on the religious background of the participants as Muslims and that of their oppressers as Hindus, and also the fact observed by Namboodiripad, that in his ignorance the Mappila could see only a particular *janmi* or a particular official as his oppresser, but not the *janmi* system or the bureaucracy which supported it (Namboodiripad, 1943:1).

The consequent limitations of the struggles were the lack of proper leadership, ideology and mobilisation, the jacquerie-like primary motive, viz., 'to gain immediate relief from pent up frustration through destruction of property and the commission of violence against persons' (see Landsberger, 1974:21; Hilton, 1974:85), restriction of the demands to only redressal of the immediate grievances, and the inadequacy of the resultant land reforms.

Despite these limitations, the uniqueness of the struggles was that they were the most protracted, the initial voice o protest against the oppression of the peasantry by the *janmis* and the earliest expressions of the pressures from 'below' foı changes in agrarian relations through land reforms.

In contrast to these struggles, those in the second phase were by the upper Hindu tenantry. The emergence of these struggles were the result of the spread of English education in Malabar and the rise of a professional middle class among the Nayars consisting of government servants, intellectuals, journalists, lawyers, and so on. This class took upon itself the leadership of the struggles as one of the expressions of their growing disenchantment with and challenge to the socio-economic supremacy of the Nambutiris. These struggles were again concentrated in south Malabar because of the concentration of the *kanam* tenants and the newly educated Nayar middle class in that region.

In keeping with the socio-professional background of the leadership, these struggles were relatively more effective and were through the idiom of what may be called the rising generation of 'intelligentsia' involving the use of press, platforms, deputations, the newly constituted legislative council and the then emerging Congress leadership.

The effectiveness of the struggles was evident from the first ever counter-mobilisation efforts of the *janmis* mainly through meetings, deputations and the legislative council, and from the fact that in less than a decade of the total duration of the active phase of these struggles the government conceded to most of the demands. But as the demands projected were mostly of the *kanam* tenants, the beneficiaries of the resultant land reforms were mostly this upper stratum of tenantry and not the vast masses of the lower tenantry. Despite this, these struggles were also unique inasmuch as they were the first organised struggles, that too by a professional middle class whose interests in *kanam* right was only incidental, and these were the struggles which gave the Kerala peasants the first lessons in class unity and organisation.

Unlike the struggles in the first two phases, those in the third phase were mostly by the lower Hindu tenantry, especially the vast masses of the lower castes and, for a long time, were concentrated in north Malabar. Their emergence was bound up with a number of factors, such as, the rise in the tempo of the nationalist movement since the 1930s, its concentration in north Malabar, especially of its socialist cadres from whom emerged the militant leadership of the struggles, the deterioration in the economic condition of the peasantry, and finally, the emergence and growth of the communist movement.

The persistence of these struggles for about four decades securing to the peasantry more and more concessions from the *janmis* and eventually the land ownership itself was mainly because of the painstaking organisational efforts of the initial years leading to the gargantuan growth of the *Karshaka Sangham* as a well-structured and broad-based mass organisation. It was also bound up with the anti-*janmi* and anti-government nature of the struggles through a wide array of effective and appealing methods. These included mass meetings, processions, conferences, study classes, social boycott of the *janmis* and the government, defiance of customary authority, petitions, inter-caste dining, dramas, hunger-songs, hunger-*jathas*, and occasionally even direct action.

With the emergence of the Communist Party, these struggles became stronger, more militant and clear-cut in their objectives. With the formation of the Kerala state and the emergence of the Communist Party as the only major political force earnestly and effectively championing the cause of the peasantry they were further consolidated and politicised, and assumed a state-wide character. When the Communist Party rose to power as the only party committed to the abolition of landlordism and the peasantry stood by it, for the first time in the history of agrarian relations in Kerala (or for that matter in the whole of India), the landlords were faced with a government which was not on their side. This, and the role of the Congress Party as the guardian of the landed interests, transformed the struggles in the remaining period of about a decade into a class war. In this war the Communist Party and the *Karshaka Sangham* counted mainly on the class identity and solidarity of the peasantry, whereas the landed interests and the Congress Party exploited issues which are of greater appeal in a traditional society to even the oppressed and exploited groups, viz., the caste, communal, cultural, and more importantly the anti-communist sentiments and prejudices.

The process by which the peasantry eventually emerged victorious, albeit in a limited sense, and the nature of their victory point out at least two issues. First, for the introduction of land reforms (or for that matter any other reforms intended to benefit the underprivileged), their organised and assiduous pressures are necessary conditions. Second, even when these conditions are created, so long as the states have only limited legislative powers and the state power is concentrated at a centre of the landed interests, the reforms necessitated by organised demands can only be a limited success.

But, even as a limited success, if the Kerala Land Reforms (Amendment) Act of 1969 (which effectively abolished traditional landlordism) is viewed as the culmination of the peasant struggles of nearly a century and a half, its importance is both historical and contemporary. Its historical importance stems from the fact that it has not only dealt the death blow to traditional landlordism in Kerala, the main source of oppression and exploitation in traditional Kerala society, but also brought to the logical conclusion a century-old controversy on the rights of the *janmis*, a controversy which became very heated after Logan submitted his report on Malabar land tenures towards the close of the nineteenth century.

Its contemporary importance is related to the recent recognition of land reforms as a prerequisite to growth with social justice and induced social change in rural areas. An act of this kind resulting in the abolition of a vital institution (such as landlordism) which reigned supreme in Kerala society for several centuries, and in the large-scale redistribution of land ownership should have had tremendous effects on the society. Whether it has had such effects is a moot issue.

Hypothetically, the effects should have been incomparably greater had the Act been effectively implemented a century or even half a century ago when the institution of landlordism was relatively more pristine. When the institution was eventually abolished it was only in a truncated form, having had already shed many of its traditional attributes because of the numerous peasant struggles. Consequently, for a full measure of the effects, the institution has to be seen in its pristine form and the effects in their historical totality. The latter, however, is very difficult because of the wide diffusion of the changes over several years.

Even assuming the effects as only the latest phase of a long process of change in terms of the almost absolute structural transformation brought about by the massive land transfers and some of the resultant changes in the land-based social order, these are certainly very great. It is, however, important to note that their magnitude has not been really felt by the people. To note some of the reasons for this failure, Mahakavi G. Sankara Kurup, one of the greatest poets of humanism and the first Jnanpeeth Award winner, wrote: 'I often feel that our Land Reforms Legislation, with its painless, yet levelling approach geared to distributive justice is not appreciated in its far-reaching implications by many, merely because we have been conditioned to the theory of blood-soaked revolutions' (see *Dharani*, November 1976:37).

N.V. Krishna Warrier, another great poet, essayist, critic, journalist and editor attributed it to two reasons. According to him the most important reason is that an overwhelming majority of the landless labourers have not benefited by the land redistribution. Related to his view is the conceptualisation of the slogan 'land to the tiller'. In the Kerala context there was not much clarity or consensus on the meaning of this slogan. A tiller could be either a hired agricultural labourer tilling someone's land or a self-cultivator tilling his own land, or both. This was especially so in pre-reforms Kerala where 'the border line between the agricultural labour class and the marginal peasantry, roughly 60 per cent of the owner cultivators [was] an amorphous one' (Lieten, 1982:68). Commenting on the KARB, the Malayalam daily *Mathrubhumi* in an editorial had pointed out that the slogan does not and will not mean anything more than the cultivator's right to cultivate (*Mathrubhumi*, 12 June 1959). K.R. Gowri's usage in Malayalam when she introduced the KARB in the Kerala assembly was *krishibhoomi krishikkaranu*, meaning farm-land to the farmer (*Mathrubhumi*, 22 December 1957).

In this connection it needs to be mentioned that in pre-reforms Kerala landless tillers were of two broad categories: those without any leased-in land who worked as hired agricultural labourers; and those who had small parcels of leased-in land. The majority of those who have benefited by the land transfers under the first scheme of the 1969 Act conferring ownership rights on the cultivating tenants belonged to the second category. By sheer economic necessity many of them might have also worked as hired agricultural labourers as they may be doing even now.

By further measures—such as, redefinition of cultivators excluding the supervisory cultivators from the benefits of the Act, rigorous implementation of the ceiling provisions, and taking over the lands of those who have principal occupations with adequate income in the non-agricultural sector—the government could have certainly managed to redistribute some more land. But despite all such measures, it is still doubtful if the 1969 Act would have solved Kerala's land problem. And considering the rate at which the price of land has been shooting up, those who now remain landless or land-poor may not find a chance in the future for acquiring any land, unless of course god Parasurama, the mythical creator of Kerala, throws his axe once again, as someone remarked, and reclaims some more land from the sea. It is in this context that the exclusion of the vast areas of plantations from the successive land reform enactments has to be seen as a major limiting factor of Kerala land reforms, making the effects of even the most thorough-going Act only partially felt.

To note Warrier's second view:

> Even those who are benefited by it [the land redistribution] find in cultivation only a dubious means of livelihood. They have of course got in their hands the records establishing their ownership in the piece of land they till, and this certainly should give a lot of psychological satisfaction to them. But this is not enough to make agriculture economically viable. Modern agriculture, as everyone knows, requires a lot of inputs and technical knowhow which are beyond the reach of an impoverished peasantry. Agriculture in Kerala cannot be sustained without the introduction of some such system as the mutual aid teams or cooperativisation and ultimately collectivisation. (See *Dharani*, November 1976:29)

While land reforms are central to any major transformation in agrarian society they alone are not sufficient. They have to be followed by other related measures. This is especially so in Kerala where agriculture is only part of a larger economy, dominated by the plantation and tertiary sectors. These follow-up measures could have been effectively introduced only if the left also had the state-power in a real socio-economic and political sense. In the absence of such power, while the left in Kerala could not bring about a larger transformation it certainly could not postpone its demands on and programmes for land reforms either. Therein lie both the strength and weakness of the left in the country—basically a dilemma of its uneven development.

# References


**Achutan, K.R.** 1971. *C. Krishnan (Jeevacharitram)*, National Book Stall, Kottayam.

**Aiyappan, A.** 1965. *Social Revolution in a Kerala Village*, Asia Publishing House, Bombay.

**Alexandrov, Yu G.** 1974. 'The Peasant Movement of Developing Countries in Asia and North Africa After the Second World War,' in Henry A. Landsberger (ed.), *Rural Protest: Peasant Movement and Social Change*, The Macmillan Press Ltd., London.

**Andalat**, (ed.) 1978. *Vegam Pora: Collected Works of C.H. Kanaran*, Chinta Publishers, Trivandrum.

————. 1979. *Sakhakkale Munnottu: Collected Works of P. Krishna Pilla*, Chinta Publishers, Trivandrum.

**Balaram, N.E.** 1973. *Keralathile Communist Prasthanam: Adyanalukalilute*, Prabhat Book House, Trivandrum.

**Bell, Clive.** 1974. 'Ideology and Economic Interests in Indian Land Reforms,' in David Lehmann (ed.), *Agrarian Reforms and Agrarian Reformism*, Faber and Faber Ltd., London.

**Bergmann, Theodor** . 1984. *Agrarian Reform in India: With Special Reference to Kerala, Karnataka, Andhra Pradesh, and West Bengal*, Agricole Publishing Company, New Delhi.

**Beteille, Andre.** 1974. *Studies in Agrarian Social Structure*, Oxford University Press, New Delhi.

**Bharateeyan, Vishnu V.M.** 1981. *Adimakalengane Udamakalaye*, Prabhat Book House, Trivandrum.

**Buchanan, Francis.** 1807. *A Journey from Madras through the Countries of Mysore, Canara, and Malabar*, 3 Vols, London.

**Byres, T.J.** 1974. 'Land Reforms, Industrialisation and the Marketed Surplus in India: An Essay on the Power of Rural Bias,' in David Lehmann (ed.), *Agrarian Reforms and Agrarian Reformism*, Faber and Faber Ltd., London.

**Choudhary, Sukhbir**. 1977. *Moplah Uprising (1921–23)*, Agam Prakashan, New Delhi.

**Dale, Stephen Frederic**. 1980. *Islamic Society on the South Asian Frontier: The Mappilas of Malabar: 1498–1922*, Clarendon Press, Oxford.

**Dames, Mansel Longsworth**, (ed.). 1921. *The Book of Duarte Barbosa: An Account of the Countries Bordering on the Indian Ocean and Their Inhabitants* (written by Duarte Barbosa, and completed about the year 1518 A.D.), The Haklyut Society, London.

**Dantwala, M.L.** 1957. 'Problems and Prospects of Land Reforms in India,' *Economic Development and Cultural Change*, 6(1), October.

**Dhanagare, D.N.** 1983. *Peasant Movements in India: 1920–1950*, Oxford University Press, Bombay.

**Dorner, Peter**. 1972. *Land Reform and Economic Development*, Penguin, Harmondsworth.

**Fic, Victor M.** 1970. *Kerala: Yenan of India: Rise of Communist Power, 1937–1969*, Nachiketa Publications, Bombay.

**Frankel, Francine R.** 1971. *India's Green Revolution: Economic Gains and Political Costs*, Princeton University Press, New Jersey.

**Gopalan, A.K.** 1975. *Manninuventi*, Chinta Publishers, Trivandrum.

————. 1976. *In the Cause of the People*, Orient Longman, Bombay.

**Gough, Kathleen**. 1959. 'Criteria of Caste Ranking in South India,' *Man in India*, 39(2), June.

————. 1978. 'Village Politics in Kerala,' in A.R. Desai, (ed.) *Rural Sociology in India*, Popular Prakashan, Bombay (first published 1958).

**Gough, Kathleen** and **David M. Schneider**, (eds.). 1972. *Matrilineal Kinship*, A.H. Wheeler and Company, Allahabad (first published 1961).

**Hart, Henry C.** and **Ronald J. Herring**. 1977. 'Political Conditions of Land Reforms: Kerala and Maharashtra,' in Robert Eric Frykenberg (ed.), *Land Tenures and Peasant in South Asia*, Orient Longman, New Delhi.

**Herring, Ronald J.** 1980. 'Abolition of Landlordism in Kerala: A Redistribution of Privilege,' *Economic and Political Weekly*, 15(26), 28 June.

————. 1983. *Land to the Tiller: The Political Economy of Agrarian Reform in South Asia*, Oxford University Press, Delhi.

**Hilton, Rodney H.** 1974. 'Peasant Society, Peasant Movements and Feudalism in Medieval Europe,' in Henry A. Landsberger (ed.), *Rural Protest: Peasant Movement and Social Change*, Macmillan Press Ltd., London.

**Houtart, Francois**, and **Geneieve Lemercinier.** 1978. 'Socio-Religious Movements in Kerala: A Reaction to the Capitalist Mode of Production,' *Social Scientist*, 6(11–12), June-July.

**India, Government of.** 1981. *Rural Labour Enquiry, 1974–75: Final Report on Employment and Unemployment of Rural Labour Households* (Part I).

————. 1986. *National Sample Survey Organisation*, 37th Round, January-December 1982, No. 330, Report on Land Holdings 1 July.

**Innes, C.A.** 1951. *Malabar*, Government Press, Madras (first published 1909).

**Iyer, Krishna L.A.** 1968. *Social History of Kerala*, Vol. I, Book Centre Publications, Madras.

————. 1970. *Social History of Kerala*, Vol. II, Book Centre Publications, Madras.

**Jeffrey, Robin.** 1976. *The Decline of Nayar Dominance: Society and Politics in Travancore, 1847–1908*, Vikas Publishing House, New Delhi.

————. 1978a. 'Peasant Movements and the Communist Party in Kerala: 1937–1957,' in D.B. Miller (ed.), *Peasants and Politics:Grass Roots Reactions to Change in Asia*, Edward, Melbourne.

————. 1978b. 'Matriliny, Marxism and the Birth of the Communist Party in Kerala, 1930–40,' *Journal of Asian Studies* 38(1), November.

**Joshi, P.C.** 1976. *Land Reforms in India: Trends and Perspectives*, Allied Publishers, New Delhi.

**Kerala, Government of.** 1968. Bureau of Economics and Statistics, *Land Reforms Survey, 1966–67, Report*, Trivandrum.

————. 1977. *Kerala Land Reforms Manual*, 2 Vols, Trivandrum.

————. *Administration Reports of the Kerala Land Board and Land Tribunals, 1970–71 to 1981–82.*

**Karat, Prakash.** 1973. 'Agrarian Relations in Malabar, 1925–1948,' *Social Scientist*, 2(2–3), September-October.

————. 1976. 'Peasant Movement in Malabar—1934–40,' *Social Scientist*, 5(2), September.

————. 1977. 'Organised Struggles of Malabar Peasantry, 1934–1940,' *Social Scientist*, 5(8), March.

**Kareem, C.K.** 1973. *Kerala Under Haider Ali and Tippu Sultan*, Kerala History Association, Cochin.

**Koshy, V.C.** 1974. 'Land Reforms in India Under the Plans,' *Social Scientist*, 2(12), July.

————. 1976. 'Land Reforms and Rural Poor,' *Dharani*, Journal of the Kerala Land Board, November.

**Kotovsky, Grigory.** 1964. *Agrarian Reforms in India*, Peoples Publishing House, New Delhi.

**Krishnaji, N.** 1979. 'Agrarian Relations and the Left Movement in Kerala—A Note on Recent Trends,' *Economic and Political Weekly*, 15(9), 3 March.

**Kunhikrishnan,** V.V. 1985. 'Tenancy Legislation in Malabar, 1880–1970: An Historical Analysis,' Ph.D. Thesis, University of Calicut.

**Kunju, Ibrahim A.P.** 1975. *Mysore-Kerala Relations in the Eighteenth Century*, Kerala Historical Society, Trivandrum.

**Kurup, K.K.N.** 1978. *The Kayyur Riot: A Terrorist Episode in the Nationalist Movement in Kerala*, Sandhya Publications, Calicut.

———. 1981. *William Logan: A Study in the Agrarian Relations of Malabar*, Sandhya Publications, Calicut.

———. 1985. *A.C. Kannan Nair: Oru Patanam*, Kerala Bhasha Institute, Trivandrum.

**Kurup, M.N.** 1982. *Kunhambuvinte Katha*, Vidyarthi Mitram Book Depot, Trivandrum.

**Ladejinsky, Wolf.** 1977. *Agrarian Reforms as Unfinished Business: The Selected Papers of Wolf Ladejinsky*, Louis J. Walinsky, (ed.), Oxford University Press, New York.

**Landsberger, Henry A.** (ed.), 1974. *Rural Protest: Peasant Movement and Social Change*, The Macmillan Press Ltd., London.

**Lee, Samuel.** 1829. *Travels of Ibn Batuta*, J. Murray, London.

**Lieten, Georges Kristoffel.** 1977. 'Education, Ideology and Politics in Kerala, 1957–59,' *Social Scientist*, 6(2), September.

———. 1982. *The First Communist Ministry in Kerala, 1957–59*, K.P. Bagchi & Company, Calcutta.

**Logan, William.** 1951a. *Malabar*, 3 Vols, Government Press, Madras (first published 1887).

———. 1951b. *A Collection of Treaties, Engagements and Other Papers of Importance Relating to British Affairs in Malabar*, Government Press, Madras.

**Madhavan, K.** 1987. *Payaswiniyude Theerathu*, Prabhat Book Depot, Trivandrum.

**Madras, Government of.** 1874. *Report of the Census of the Madras Presidency*, Vol. 1, Government Gazette Press, Madras.

———. 1874. *Census of the Madras Presidency: Statement of Population of 1871 in each Village of the Malabar District*, Government Gazette Press, Madras.

———. 1882. *Malabar Special Commission, 1881–82, Malabar Land Tenures*, 2 Vols, Madras.

———. 1885. *Correspondence on Malabar Land Tenures*, Government Press, Madras.

———. 1887. *Report with Appendices of the Malabar Land Tenures Committee, 1885*, Government Press, Madras.

———. 1928. *Malabar Tenancy Committee, Report, 1927–28*, Vol. 1, Madras, Government Press.

———. 1940. *Report of the Malabar Tenancy Committee, 1939*, Vol. 1, Government Press, Madras.

———. *Fortnightly Reports* for the relevant years.

**Malaviya, H.D.** 1959. *Kerala: A Report to the Nation*. Peoples Publishing House, New Delhi.

**Mayer, Adrian C.** 1952. *Land and Society in Malabar*, Oxford University Press, Bombay.

**Mazumdar, R.C.** 1973. In *Smaranika*, Mappila Rebellion Martyrs Convention Committee Publication, Calicut.

**Mencher, Joan P.** 1966. 'Namboodiri Brahmins: An Analysis of a Traditional Elite in Kerala,' *Journal of Asian and African Studies*, 1.

———. 1966. 'Kerala and Madras: A Comparative Study of Ecology and Social Structure,' *Ethnology*, 5(2), April.

———. 1966. 'The Namboodiri Brahmins of Kerala,' *Natural History*, LXXV, May.

———. 1976. 'Land Reform and Socialism: The Case of Kerala,' in S. Devadas Pillai (ed.), *Aspects of Changing India: Studies in Honour of Professor G.S. Ghurye*, Popular Prakashan, Bombay.

———. 1978. 'Agrarian Relations in Two Rice Regions of Kerala,' *Economic and Political Weekly*, 12(6–7), February.

———. 1980. 'The Lessons and Non-Lessons of Kerala: Agricultural Labourers and Poverty,' *Economic and Political Weekly*, 15(41–43).

**Menon, C. Achutha.** 1969. *What happened in Kerala: Review of the 30 months of Namboodiripad Government*, Communist Party Publication, New Delhi.

———. 1969. *Kisan Patapusthakam*, Mathrubhumi Printing & Publishing Co. Ltd., Calicut.

**Menon, Kesava K.P.** 1976. *Kazhinha Kalam*, Mathrubhumi Printing & Publishing Co. Ltd., Calicut.

**Menon, Padmanabha K.P.** 1924–37. *History of Kerala*, 4 Vols, Cochin Government Press, Ernakulam.

**Menon, Sreedhara A.** 1964. *A Summary of Kerala History*, National Book Stall, Kottayam.

———. 1978. *Cultural Heritage of Kerala: An Introduction*, East-West Publications, Cochin.

———. 1979. *Social and Cultural History of Kerala*, Sterling Publishers, New Delhi.

**Menon, V.R.** 1973. *Mathrubhumiyute Charitram*, Vol. 1, Mathrubhumi Printing & Publishing Co. P. Ltd., Calicut.

**Miller, R.E.** 1976. *Mappila Muslims of Kerala: A Study in Islamic Trends*, Orient Longman, Madras.

**Nair, Madhavan K.** 1987. *Malabar Kalapam*, Mathrubhumi Printing & Publishing Co. Ltd., Calicut.

**Nair, Narayanan P.** 1973. *Aranootantilute*, National Book Stall, Kottayam.

**Nair, Sankaran G.**, 1971. *Kudiyan Prasthanam Malabaril*, Malayalam Encyclopaedia, Vol. 4, National Book Stall, Kottayam.

**Namboodiripad, E.M.S.** 1943. *A Short History of the Peasant Movement in Kerala*, Peoples Publishing House, Bombay.

**Namboodiripad, E.M.S.** 1967. *Kerala, Yesterday, Today, and Tomorrow*, National Book Agency, Calcutta.

————. 1976. *How I Became A Communist*, Chinta Publishers, Trivandrum.

————. 1981. *Crisis into Chaos: Political India 1981*, Orient Longman, New Delhi.

**Nayanar, E.K.** 1982. *My Struggles: An Autobiography*, Vikas Publishing House, Delhi.

**Neale, Walter C.** 1962. *Economic Change in Rural India: Land Tenure and Reforms in Uttar Pradesh: 1800–1955*, Kennikat Press, London.

**Nossiter, T.J.** 1982. *Communism in Kerala: A Study in Political Adaptation*, Oxford University Press, Delhi.

**Oommen, M.A.** 1975. *A Study on Land Reforms in Kerala*, Oxford & IBH Publishing Co., New Delhi.

**Oommen, T.K.** 1985. *From Mobilization to Institutionalization: The Dynamics of Agrarian Movement in Twentieth Century Kerala*, Popular Prakashan, Bombay.

**Panikkar, Gopala T.K.** 1901. *Malabar and its Folk*, C.A. Natesan and Co., Madras.

**Panikkar, K.N.**, 1978. 'Agrarian Legislations and Social Classes: A Case Study of Malabar,' *Economic and Political Weekly*, 7(21), 27 May.

————. 1979. 'Peasant Revolts in Malabar in the Nineteenth and Twentieth Centuries,' in A.R. Desai (ed.), *Peasant Struggles in India*, Oxford University Press, Delhi.

**Pilla Kunhan Elamkulam P.N.** 1966. *Janmi Sampradayam Keralathil*, National Book Stall, Kottayam.

**Pothuval, A.K.** 1976. *Keralathile Karshaka Prasthanam*, Prabhat Book House, Trivandrum.

————. 1978. *Karshaka Samara Kadha*, Prabhat Book House, Trivandrum.

————. 1981. *Keralathil Janmitham Avasanichu*, Prabhat Book House, Trivandrum.

**Pottekkat, S.K.**, et al. 1978. *Muhammad Abdurahiman: A Political Biography*, Calicut.

**Raj, K.N.** and **Michael Tharakan.** 1984. 'Agrarian Reform in Kerala and its Impact on the Rural Economy: A Preliminary Assessment,' in Ajit Kumar Ghose (ed.), *Agrarian Reform in Contemporary Developing Countries*, Select Book Service Syndicate, New Delhi.

**Raja, Unni C.** 1974. *Karshaka Prasthanam*, Kerala Charitram (compilation of Kerala History), Vol. II, Ernakulam.

**Rao, M.S.A.** 1957. *Social Change in Malabar*, Popular Book Depot, Bombay.

————. 1979. *Social Movements and Social Transformation*, The Macmillan Co. of India Ltd., Madras.

**Rao, T.V.S.**, 1970. 'Land Legislation in Kerala, 1800–1950,' *Artha Vijnana*, 12(1–2), March-June.

**Ravindran, T.K.** 1973. In *Smaranika*, Mappila Rebellion Martyrs Convention Committee Publication, Calicut.

**Shea, Thomas W. Jr.** 1979. 'Barriers to Economic Development in Traditional Societies: Malabar—A Case Study,' *Journal of Economic History*, XIX(1), March.

**Tharakan, Michael P.K.** 1982. 'The Kerala Land Reforms (Amendment) Bill, 1979: A Note,' in Sheo Kumar Lal (ed.), *Sociological Perspectives of Land Reforms*, Agricole Publishing Academy, New Delhi.

**Thorner, Daniel.** 1976. *The Agrarian Prospect in India*, Allied Publishers Ltd., New Delhi (first published 1956).

**Varghese, T.C.** 1970. *Agrarian Change and Economic Consequences: Land Tenures in Kerala, 1850–1960*, Allied Publishers, Calcutta.

**Varughese, K.V.** 1978. *United Front Government in Kerala: 1967–1969*, The Christian Literature Society, Madras.

**Warrier, K.V.K.** 1970. 'Kerala Shows the Way,' *Mainstream*, 8(43–45), 27 June–11 July.

**Wood, Conrad.** 1974. 'Historical Background of the Moplah Rebellion: Outbreaks, 1836–1919,' *Social Scientist*, 3(1), August.

# Index